PROBABLY OVERTHINKING IT

PROBABLY OVERTHINKING IT

HOW TO USE DATA TO ANSWER QUESTIONS,
AVOID STATISTICAL TRAPS, AND MAKE
BETTER DECISIONS

ALLEN B. DOWNEY

THE UNIVERSITY OF CHICAGO PRESS

Chicago and London

The University of Chicago Press, Chicago 60637
The University of Chicago Press, Ltd., London
© 2023 by Allen B. Downey

Published 2023
Printed in the United States of America

32 31 30 29 28 27 26 25 24 23   1 2 3 4 5

ISBN-13: 978-0-226-82258-7 (cloth)
ISBN-13: 978-0-226-82259-4 (e-book)
DOI: https://doi.org/10.7208/chicago/9780226822594.001.0001

Library of Congress Cataloging-in-Publication Data

Names: Downey, Allen, author.
Title: Probably overthinking it : how to use data to answer
questions, avoid statistical traps, and make better decisions /
Allen B. Downey.
Other titles: How to use data to answer questions, avoid statistical
traps, and make better decisions
Description: Chicago ; London : The University of Chicago Press,
2023. | Includes bibliographical references and index.
Identifiers: LCCN 2023016550 | ISBN 9780226822587 (cloth) |
ISBN 9780226822594 (ebook)
Subjects: LCSH: Statistics.
Classification: LCC QA276 .D67 2023 | DDC 519.5—dc23/eng20230710
LC record available at https://lccn.loc.gov/2023016550

♾ This paper meets the requirements of ANSI/NISO Z39.48-1992
(Permanence of Paper).

CONTENTS

INTRODUCTION

Let me start with a premise: we are better off when our decisions are guided by evidence and reason. By "evidence," I mean data that is relevant to a question. By "reason" I mean the thought processes we use to interpret evidence and make decisions. And by "better off," I mean we are more likely to accomplish what we set out to do—and more likely to avoid undesired outcomes.

Sometimes interpreting data is easy. For example, one of the reasons we know that smoking causes lung cancer is that when only 20% of the population smoked, 80% of people with lung cancer were smokers. If you are a doctor who treats patients with lung cancer, it does not take long to notice numbers like that.

But interpreting data is not always that easy. For example, in 1971 a researcher at the University of California, Berkeley, published a paper about the relationship between smoking during pregnancy, the weight of babies at birth, and mortality in the first month of life. He found that babies of mothers who smoke are lighter at birth and more likely to be classified as "low birthweight." Also, low-birthweight babies are more likely to die within a month of birth, by a factor of 22. These results were not surprising.

However, when he looked specifically at the low-birthweight babies, he found that the mortality rate for children of smokers is *lower*, by a factor of two. That *was* surprising. He also found that among low-birthweight babies, children of smokers are *less likely* to have birth defects, also by a factor of 2. These results make maternal smoking seem beneficial for low-birthweight babies, somehow protecting them from birth defects and mortality.

The paper was influential. In a 2014 retrospective in the *International Journal of Epidemiology*, one commentator suggests it was responsible for "holding up anti-smoking measures among pregnant women for perhaps a decade" in the United States. Another suggests it "postponed by several years any campaign to change mothers' smoking habits" in the United Kingdom.

But it was a mistake. In fact, maternal smoking is bad for babies, low birthweight or not. The reason for the apparent benefit is a statistical error I will explain in chapter 7.

Among epidemiologists, this example is known as the low-birthweight paradox. A related phenomenon is called the obesity paradox. Other examples in this book include Berkson's paradox and Simpson's paradox. As you might infer from the prevalence of "paradoxes," using data to answer questions can be tricky. But it is not hopeless. Once you have seen a few examples, you will start to recognize them, and you will be less likely to be fooled. And I have collected a lot of examples.

So we can use data to answer questions and resolve debates. We can also use it to make better decisions, but it is not always easy. One of the challenges is that our intuition for probability is sometimes dangerously misleading. For example, in October 2021, a guest on a well-known podcast reported with alarm that "in the [United Kingdom] 70-plus percent of the people who die now from COVID are fully vaccinated." He was correct; that number was from a report published by Public Health England, based on reliable national statistics. But his implication—that the vaccine is useless or actually harmful—is wrong.

As I'll show in chapter 9, we can use data from the same report to compute the effectiveness of the vaccine and estimate the number of lives it saved. It turns out that the vaccine was more than 80% effective at preventing death and probably saved more than 7000 lives, in a four-week period, out of a population of 48 million. If you ever find yourself with the opportunity to save 7000 people in a month, you should take it.

The error committed by this podcast guest is known as the base rate fallacy, and it is an easy mistake to make. In this book, we will see examples from medicine, criminal justice, and other domains where decisions based on probability can be a matter of health, freedom, and life.

THE GROUND RULES

Not long ago, the only statistics in newspapers were in the sports section. Now, newspapers publish articles with original research, based on data collected and analyzed by journalists, presented with well-designed, effective visualization. And data visualization has come a long way. When *USA Today* started publishing in 1982, the infographics on their front page were a novelty. But many of them presented a single statistic, or a few percentages in the form of a pie chart.

Since then, data journalists have turned up the heat. In 2015, "The Upshot," an online feature of the *New York Times,* published an interactive, three-dimensional representation of the yield curve—a notoriously difficult concept in economics. I am not sure I fully understand this figure, but I admire the effort, and I appreciate the willingness of the authors to challenge the audience. I will also challenge my audience, but I won't assume that you have prior knowledge of statistics beyond a few basics. Everything else, I'll explain as we go.

Some of the examples in this book are based on published research; others are based on my own observations and exploration of data. Rather than report results from a prior work or copy a figure, I get the data, replicate the analysis, and make the figures myself. In some cases, the original work did not hold up to scrutiny; those examples are not in the book. For some examples, I was able to repeat the analysis with more recent data. These updates are enlightening. For example, the low-birthweight paradox, which was first observed in the 1970s, persisted into the 1990s, but it has disappeared in the most recent data.

All of the work for this book is based on tools and practices of reproducible science. I wrote each chapter in a Jupyter notebook, which combines the text, computer code, and results in a single document. These documents are organized in a version-control system that helps to ensure they are consistent and correct. In total, I wrote about 6000 lines of Python code using reliable, open-source libraries like NumPy, SciPy, and pandas. Of course, it is possible that there are bugs in my code, but I have tested it to minimize the chance of errors that substantially affect the results. My Jupyter notebooks are available online so that anyone can replicate the analysis I've done with the push of a button.

With all that out of the way, let's get started.

SOURCES AND RELATED READING

The bracketed numbers in the "Sources" sections refer to numbered entries in the bibliography.

- You can view the three-dimensional visualization of the yield curve at the *New York Times* blog "The Upshot" [5].
- My Jupyter notebooks are available from GitHub [35].

CHAPTER 1

ARE YOU NORMAL? HINT: NO

What does it mean to be normal? And what does it mean to be weird? I think there are two factors that underlie our intuition for these ideas:

- "Normal" and "weird" are related to the idea of average. If by some measurement you are close to the average, you are normal; if you are far from the average, you are weird.
- "Normal" and "weird" are also related to the idea of rarity. If some ability or characteristic of yours is common, it is normal; if it's rare, it's weird.

Intuitively, most people think that these things go together; that is, we expect that measurements close to the average are common and that measurements far from average are rare. For many things, this intuition is valid. For example, the average height of adults in the United States is about 170 cm. Most people are close to this average: about 64% of adults are within 10 cm plus or minus; 93% are within 20 cm. And few people are far from average: only 1% of the population is shorter than 145 cm or taller than 195 cm.

But what is true when we consider a single characteristic turns out to be partly false when we consider a few characteristics at once, and spectacularly false when we consider more than a few. In fact, when we consider the many ways people are different, we find that

- people near the average are rare or nonexistent,
- everyone is far from average, and
- everyone is roughly the same distance from average.

At least in a mathematical sense, no one is normal, everyone is weird, and everyone is the same amount of weird. To show why this is true, we'll start with a single measurement and work our way up to hundreds, and then thousands. I'll introduce the Gaussian curve, and we'll see that it fits many of these measurements remarkably well.

In addition to physical measurements from a survey of military personnel, we'll also consider psychological measurements, using survey results that quantify the Big Five personality traits.

PRESENT . . . ARMS

How tall are you? How long are your arms? How far is it from the radiale landmark on your right elbow to the stylion landmark on your right wrist?

You might not know that last one, but the US Army does. Or rather, they know the answer for the 6068 members of the armed forces they measured at the Natick Soldier Center (just a few miles from my house) as part of the Anthropometric Surveys of 2010–2011, which is abbreviated army-style as ANSUR-II. In addition to the radiale-stylion length of each participant, the ANSUR dataset includes 93 other measurements "chosen as the most useful ones for meeting current and anticipated Army and [Marine Corps] needs." The results were declassified in 2017 and are available to download.

We will explore all of the measurements in this dataset, starting with height. The following figure shows the distribution of heights for the male and female participants in the survey. The vertical axis shows the percentage of participants whose height falls in each one-centimeter interval. Both distributions have the characteristic shape of the "bell curve."

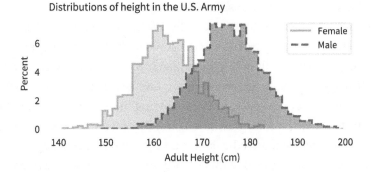

Distributions of height in the U.S. Army

One of the earliest discoveries in the history of statistics is that there is a simple model that matches the shape of these curves with remarkable accuracy. Written as a mathematical formula, it might not seem simple, depending on how you feel about the Greek alphabet. So instead I will describe it as a game. Please think of a number (but keep it small). Then follow these steps:

1. Square your number. For example, if you chose two, the result is four.
2. Take the result from the previous step and raise 10 to that power. If you chose two, the result is 10^4, which is 10,000.
3. Take that result and invert it. If you chose two, the result is 1/10,000.

Now I'll do the same calculation for a range of values from −2 to 2 and plot the results. Here's what it looks like:

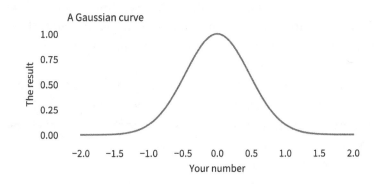

A Gaussian curve

This result is known as the Gaussian curve, after the mathematician Carl Friedrich Gauss. Actually, it's just one of many Gaussian curves. By changing the rules of the game we just played, we can shift the curve to the right or left and make it wider or narrower. And by shifting and stretching, we can often find a Gaussian curve that's a good match for real data. In the following figure, the shaded areas show the distributions of height again; the lines show the Gaussian curves I chose to match the data. And they fit the data pretty well. If you have seen results like this before, they might not surprise you; but maybe they should.

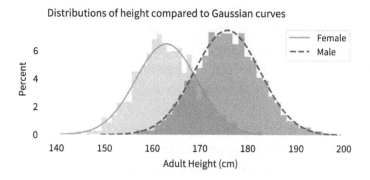

Distributions of height compared to Gaussian curves

Mathematically, the Gaussian curve is simple. The operations we used to compute it—squaring, raising to a power, and inverting numbers—are familiar to people with a basic math education. In contrast, people are complicated. Nearly everything about us is affected by hundreds or thousands of interacting causes and effects, including our genes and everything about our environment from conception to adulthood.

So if we measure a few thousand people, plot the results, and find that they fit a simple model so well, we should be surprised. Furthermore, it's not just people and it's not just height. If you choose individuals from almost any species and measure almost any part of them, the distribution of measurements will be approximately Gaussian. In many cases, the approximation is remarkably close. Naturally, you might wonder why.

The answer comes in three parts:

- Physical characteristics like height depend on many factors, both genetic and environmental.
- The contribution of these factors tends to be additive; that is, the measurement is the sum of many contributions.
- In a randomly chosen individual, the set of factors they have inherited or experienced is effectively random.

When we add up the contributions of these random factors, the resulting distribution tends to follow a Gaussian curve.

To show that that's true, I will use a model to simulate random factors, generate a set of random heights, and compare them to the measured heights of the participants in the ANSUR dataset. In my model, there are 20 factors that influence height. That's an arbitrary choice; it would also work with more or fewer. For each factor, there are two possibilities, which you can think of as different genes or different environmental conditions. To model an individual, I generate a random sequence of 0s and 1s that indicate the absence or presence of each factor. For example:

- If there are two forms of a gene (known as alleles), 0 might indicate one of the alternatives, and 1 the other.
- If a particular nutrient contributes to height at some point in development, 0 and 1 might indicate the deficiency or sufficiency of the nutrient.
- If a particular infection can detract from height, 0 and 1 might indicate the absence or presence of the infectious agent.

In the model, the contribution of each factor is a random value between −3 cm and 3 cm; that is, each factor causes a person to be taller or shorter by a few centimeters.

To simulate a population, I generate random factors (0s and 1s) for each individual, look up the contribution of each factor, and add them up. The result is a sample of simulated heights, which we can compare to the actual heights in the dataset. The following figure shows the results. The shaded areas show the distributions of the actual heights again; the lines show the distributions of the simulated heights.

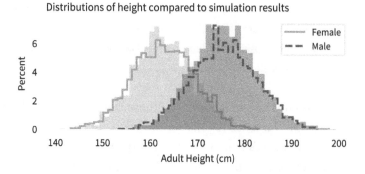

Distributions of height compared to simulation results

The results from the simulation are a good fit for the data. Before I ran this simulation, I had a pretty good idea what the results would look like because of the Central Limit Theorem, which states that the sum of a large number of random values follows a Gaussian distribution. Mathematically, the theorem is only true if the random values come from the same distribution and they are not correlated with each other.

Of course, genetic and environmental factors are more complicated than that. In reality, some contributions are bigger than others, so they don't all come from the same distribution. Also, they are likely to be correlated with each other. And their effects are not purely additive; they can interact with each other in more complicated ways.

However, even when the requirements of the Central Limit Theorem are not met exactly, the combined effect of many factors will be approximately Gaussian as long as

- none of the contributions are much bigger than the others,
- the correlations between them are not too strong, and
- the total effect is not too far from the sum of the parts.

Many natural systems satisfy these requirements, which is why so many distributions in the world are approximately Gaussian.

COMPARING DISTRIBUTIONS

So far, I have represented distributions using histograms, which show a range of possible values on the horizontal axis and percentages on the vertical axis. Before we go on, I want to present another way to visualize distributions that is particularly good for making comparisons. This visualization is based on *percentile ranks*.

If you have taken a standardized test, you are probably familiar with percentile ranks. For example, if your percentile rank is 75, that means that you did as well as or better than 75% of the people who took the test. Or if you have a child, you might be familiar with percentile ranks from a pediatric growth chart. For example, a two-year-old boy who weighs 11 kg has percentile rank 10, which means he weighs as much or more than 10% of children his age.

Computing percentile ranks is not hard. For example, if a female participant in the ANSUR survey is 160 cm tall, she is as tall or taller than 34% of the female participants, so her percentile rank is 34. If a male participant is 180 cm, he is as tall or taller than 75% of the male participants, so his percentile rank is 75. If we compute the percentile rank for each participant in the same way and plot these percentile ranks on the vertical axis, the results look like this:

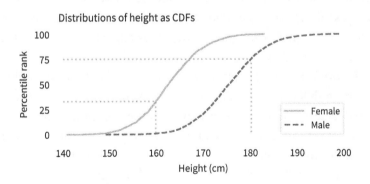

This way of representing a distribution is called a cumulative distribution function (CDF). In this example, the solid curve shows the

CDF of height for female participants; the dashed curve is the CDF for male participants. The dotted lines indicate the height and percentile rank for two examples: in the CDF for female participants, height 160 cm corresponds to percentile rank 34; in the CDF for male participants, height 180 cm corresponds to percentile rank 75.

It might not be obvious yet why this way of plotting distributions is useful. The primary reason, in my opinion, is that it provides a good way to compare distributions. For example, in the following figure, the lines show the CDFs of height again. The shaded areas show the CDFs of Gaussian distributions I chose to fit the data.

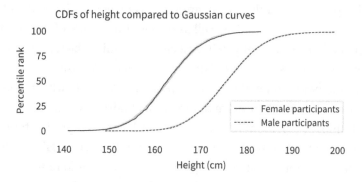

The width of the shaded areas shows how much variation we expect if we use the Gaussian model to generate simulated heights. Where a line falls inside the shaded area, the data are consistent with the Gaussian model. If it falls outside the shaded area, that indicates a deviation bigger than we would expect to see due to random variation. In these examples, both curves fall within the bounds of variation we expect.

Before we go on, I want to say something about the word "deviation" in this context, which is used to mean a difference between the data and the model. There are two ways to think about deviation. One, which is widespread in the history of statistics and natural philosophy, is that the model represents some kind of ideal, and if the world fails to meet this ideal, the fault lies in the world, not the model.

In my opinion, this is nonsense. The world is complicated. Sometimes we can describe it with simple models, and it is often useful when we can. And sometimes we can find a reason the model fits the data, which can help explain why the world is as it is. But when the world deviates from the model, that's a problem for the model, not a deficiency of the world.

Having said all that, let's see how big those deviations are.

HOW GAUSSIAN IS IT?

Out of 94 measurements in the ANSUR dataset, 93 are well modeled by a Gaussian distribution; one is not. In this chapter we'll explore the well-behaved measurements. In chapter 4 I'll reveal the exception.

For each measurement, I chose the Gaussian distribution that best fits the data and computed the maximum vertical distance between the CDF of the data and the CDF of the model. Using the size of this deviation, I identified the measurements that are the most and the least Gaussian. For many of the measurements, the distributions are substantially different for men and women, so I considered measurements from male and female participants separately.

Of all measurements, the one that is the closest match to the Gaussian distribution is the popliteal height of the male participants, which is the "vertical distance from a footrest to the back of the right knee." The following figure shows the distribution of these measurements as a dashed line and the Gaussian model as a shaded area. To quantify how well the model fits the data, I computed the maximum vertical distance between them; in this example, it is 0.8 percentile ranks, at the location indicated by the vertical dotted line. The deviation is not easily discernible.

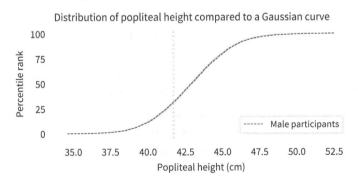

Distribution of popliteal height compared to a Gaussian curve

Of the well-behaved measurements, the one that is the worst match to the Gaussian model is the forearm length of the female participants, which is the distance I mentioned earlier between the radiale landmark on the right elbow and the stylion landmark on the right wrist. The following figure shows the distribution of these measurements and a Gaussian model. The maximum vertical distance between them is 4.2 percentile ranks, at the location indicated by the vertical dotted line; it looks like there are more measurements between 24 and 25 cm than we would expect in a Gaussian distribution. Even so, the deviation is small, and for many purposes, the Gaussian model would be good enough.

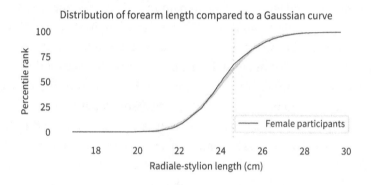

Distribution of forearm length compared to a Gaussian curve

THE MYTH OF THE "AVERAGE MAN"

We've seen that variation in physical characteristics is often well modeled by a Gaussian distribution. A characteristic of these distributions is that most people are close to the mean, with fewer and fewer people far from the mean in either direction. But as I said in the introduction of this chapter, what is true when we consider a single characteristic turns out to be counterintuitively false when we consider a few characteristics at once, and spectacularly false when we consider more than a few. In particular, when we consider the many ways each individual differs from the average, we find that people close to average in every way are rare or nonexistent.

This observation was made most famously by Gilbert Daniels in a technical report he wrote for the US Air Force in 1952, with the title "The 'Average Man'?" In the introduction he explains that

the tendency to think in terms of the "average man" is a pitfall into which many persons blunder when attempting to apply human body size data to design problems. Actually, it is virtually impossible to find an "average man" in the Air Force population.

As evidence, he uses data from the Air Force Anthropometric Survey of 1950, a precursor of the ANSUR dataset we've used in this chapter. This dataset includes 131 measurements from 4063 Air Force "flying personnel," all male. From these, Daniels selects 10 measurements "useful in clothing design." Coyly, he mentions that we would get similar results if we chose measurements useful in airplane cockpit design, which was the actual but unstated purpose of the report.

Daniels finds that of the 4063 men, 1055 are of "approximately average" height, which he defines to be within the "middle 30% of the total population." Of those, 302 are of approximately average chest circumference. Of those, 143 are of approximately average sleeve length. He continues, filtering out anyone who falls outside the middle 30% of the population by any of the other measurements. In the end, he finds three who make it past the first 8 measurements, two who make it past the first 9, and zero who were approximately average on all 10. If a uniform—or cockpit—was designed to fit the average man on all 10 measurements, it would fit no one.

Daniels suggests that "conclusions similar to those reported here would have been reached if the same type of analysis had been applied to body size data based on almost any group of people." To see if he's right, we'll replicate his analysis using data from the participants in the ANSUR survey. Many of the measurements in the ANSUR survey are the same as in the Air Force survey, but not all. Of the 10 Daniels chose, I found eight identical measurements in the ANSUR dataset and two replacements that are similar. The following table shows the names of these measurements, the mean and standard deviations of the values, the low and high ends of the range considered "approximately average," and the percentage of survey participants who fall in the range.

Measurement (cm)	Mean	Std dev	Low	High	Percentage in range
Stature (height)	175.6	6.9	173.6	177.7	23.2
Chest circumference	105.9	8.7	103.2	108.5	22.9
Sleeve length	89.6	4.0	88.4	90.8	23.1
Crotch height	84.6	4.6	83.2	86.0	22.1
Vertical trunk circ.	166.5	9.0	163.8	169.2	24.2
Hip breadth sitting	37.9	3.0	37.0	38.8	24.8
Neck circumference	39.8	2.6	39.0	40.5	25.2
Waist circumference	94.1	11.2	90.7	97.4	22.1
Thigh circumference	62.5	5.8	60.8	64.3	24.9
Crotch length	35.6	2.9	34.7	36.5	22.1

You might notice that these percentages are lower than the 30% Daniels claims. In an appendix, he explains that he used "a range of plus or minus three-tenths of a standard deviation" from the mean. He chose this range because it seemed "reasonable" to him and because it is the "equivalent of a full clothing size." To reproduce his analysis, I followed the specification in the appendix.

Now we can replicate Daniels's analysis using the measurements as "hurdles in a step-by-step elimination," starting from the top of the previous table and working down. Of 4086 male participants, 949 are approximately average in height. Of those, 244 are approximately average in chest circumference. Of those, 87 are approximately average in sleeve length.

Approaching the finish line, we find three participants that make it past 8 "hurdles," two that make it past 9, and zero that make it past all 10. Remarkably, the results from the last three hurdles are identical to the results from Daniels's report. In the ANSUR dataset, if you design for the "average man," you design for no one.

The same is true if you design for the average woman. The ANSUR dataset includes fewer women than men: 1986 compared to 4086. So we can use a more generous definition of "approximately average," including anyone within 0.4 standard deviations of the mean. Even so, we find only two women who make it past the first 8 hurdles, one who makes it past 9, and none who make it past all 10.

So we can confirm Daniels's conclusion (with a small update): "The 'average [person]' is a misleading and illusory concept as a basis for design criteria, and it is particularly so when more than one dimension is being considered." And this is not only true for physical measurements, as we'll see in the next section.

THE BIG FIVE

Measuring physical characteristics is relatively easy; measuring psychological characteristics is more difficult. However, starting in the 1980s, psychologists developed a taxonomy based on five personality traits and surveys that measure them. These traits, known as the "Big Five," are extroversion, emotional stability, agreeableness, conscientiousness, and openness to experience. Emotional stability is sometimes reported on a reversed scale as "neuroticism," so high emotional stability corresponds to low neuroticism. In psychological literature, "extroversion" is often written as "extraversion," but I will use the common spelling.

These traits were initially proposed, and gradually refined, with the goal of satisfying these requirements:

- They should be interpretable in the sense that they align with characteristics people recognize in themselves and others. For example, many people know what extroversion is and have a sense of how they compare to others on this scale.
- They should be stable in the sense that they are mostly unchanged over a person's adult life, and consistent in the sense that measurements based on different surveys yield similar results.
- The traits should be mostly uncorrelated with each other, which indicates that they actually measure five different things, not different combinations of a smaller number of things.
- They do not have to be complete; that is, there can be (and certainly are) other traits that are not measured by surveys of the Big Five.

At this point, the Big Five personality traits have been studied extensively and found to have these properties, at least to the degree we can reasonably hope for.

You can take a version of the Big Five Personality Test, called the International Personality Item Pool (IPIP) online. This test is run by the Open-Source Psychometrics Project, which "exists for the promotion of open source assessments and open data." In 2018 they published anonymous responses from more than one million people who took the test and agreed to make their data available for research. Some people who took the survey did not answer all of the questions; if we drop them, we have data from 873,173 who completed the survey. It is possible that dropping incomplete tests will disproportionately exclude people with low conscientiousness, but let's keep it simple.

The survey consists of 50 questions, with 10 questions intended to measure each of the five personality traits. For example, one of the prompts related to extroversion is "I feel comfortable around people"; one of the prompts related to emotional stability is "I am relaxed most of the time." People respond on a five-point scale from "strongly disagree" to "strongly agree." I scored the responses like this:

- "Strongly agree" scores 2 points.
- "Agree" scores 1 point.
- "Neutral" scores 0 points.
- "Disagree" scores −1 point.
- "Strongly disagree" scores −2 points.

For some questions, the scale is reversed; for example, if someone strongly agrees that they are "quiet around strangers," that counts as −2 points on the extroversion score.

Since there are 10 questions for each trait, the maximum score is 20 and the minimum score is −20. For each of the five traits, the following figure shows the distributions of total scores for more than 800,000 respondents.

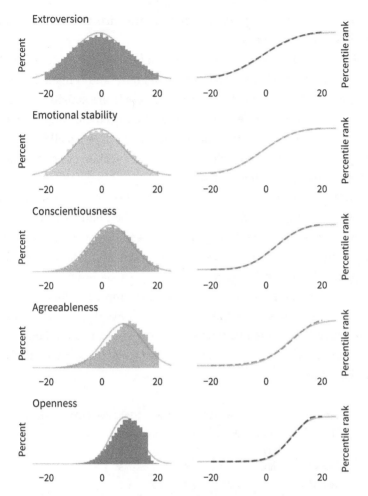

The figures on the left show histograms; the figures on the right show CDFs. In both sets of figures, the shaded line is a Gaussian distribution I chose to fit the data. The Gaussian model fits the first three distributions well (extroversion, emotional stability, and conscientiousness) except in the extreme tails.

The other two distributions are skewed to the left; that is, their tails extend farther left than right. It's possible that these deviations from the Gaussian model are the result of measurement error. In particular, agreeableness and openness might be subject to social desirability bias, which is the tendency of survey respondents to shade their answers to questions like these in the direction they know is more socially acceptable.

For example, two of the prompts related to agreeableness are "I am not interested in other people's problems," and "I insult people." It's possible that even callous, abrasive people are sensitive enough to avoid admitting these faults. Similarly, two of the prompts related to openness are "I use difficult words," and "I have excellent ideas." It's possible that some people with a limited vocabulary are mistaken about the quality of their ideas.

On the other hand, it seems like at least some unconscientious people are willing to admit that "I leave my belongings around" and "I shirk my duties." So it may be that these measurements reflect actual distributions of these attributes in the population, and it just happens that those distributions are not very Gaussian.

Regardless, let's see what happens if we apply Daniels's analysis to the Big Five data. The following table shows the mean and standard deviation of the five scores, the range of values we'll consider "approximately average," and the percentage of the sample that falls in that range.

Trait	Mean	Std dev	Low	High	Percentage in range
Extroversion	−0.4	9.1	−3.1	2.3	23.4
Emotional stability	−0.7	8.6	−3.2	1.9	20.9
Conscientiousness	3.7	7.4	1.4	5.9	20.2
Agreeableness	7.7	7.3	5.5	9.9	21.1
Openness	8.5	5.2	7.0	10.1	28.3

For each trait, the "average" range contains 20–28% of the population. Now if we treat each trait as a hurdle and select people who are close to average on each one, the following table shows the results. The first column shows the number of people who make it past each hurdle; the second column shows the percentages.

Trait	Count	Percentage
Extroversion	204,077	23.4
Emotional stability	46,988	5.4
Conscientiousness	10,976	1.3
Agreeableness	2981	0.3
Openness	926	0.1

Of the 873,173 people we started with, about 204,000 are close to the average in extroversion. Of those, about 47,000 are close to the average in emotional stability. And so on, until we find 926 who are close to average on all five traits, which is barely one person in a thousand. As Daniels showed with physical measurements, we have shown with psychological measurements: when we consider more than a few dimensions, people who are close to the average are rare or nonexistent. In fact, nearly everyone is far from average by some measure.

But that's not all. If we consider a large number of measurements, it's not just that everyone is far from average; it turns out that everyone is approximately the same distance from average. We'll see why in the next section.

WE ARE ALL EQUALLY WEIRD

How weird are you? One way to quantify that is to count the number of measurements where you are far from average. For example, using the Big Five data again, I counted the number of traits where each respondent falls outside the range we defined as "approximately average." We can think of the result as a kind of "weirdness score," where 5 means they are far from average on all five traits, and 0 means they are far from average on none. The following figure shows the distribution of these scores for the roughly 800,000 people who completed the Big Five survey.

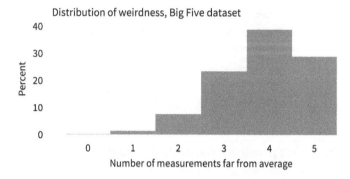

Distribution of weirdness, Big Five dataset

As we've already seen, very few people are close to average on all five traits. Almost everyone is weird in two or more ways, and the majority (68%) are weird in four or five ways!

The distribution of weirdness is similar with physical traits. Using the 93 measurements in the ANSUR dataset, we can count the number of ways each participant deviates from average. The following figure shows the distribution of these counts for the male ANSUR participants.

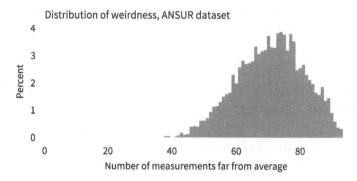

Distribution of weirdness, ANSUR dataset

Nearly everyone in this dataset is "weird" in at least 40 ways, and 90% of them are weird in at least 57 ways. With enough measurements, being weird is normal. In fact, as the number of measurements increases, the width of this distribution gets narrower; that is, the difference between the most normal person and the weirdest gets smaller.

To demonstrate, I'll use the ANSUR measurements to compute all possible ratios of two measurements. Some of these ratios are more meaningful than others, but at least some of them are features that affect clothing design (like the ratio of waist circumference to chest circumference), cockpit design (like the ratio of arm length to leg length), or perceived attractiveness (like the ratio of face width to face height).

With 93 measurements and 4278 ratios, there are a total of 4317 ways to be weird. The following figure shows the distribution of weirdness scores for the male participants. With these measurements, all participants fall in a relatively narrow range of weirdness. The most "normal" participant deviates from average in 2446 ways; the weirdest in 4038 ways.

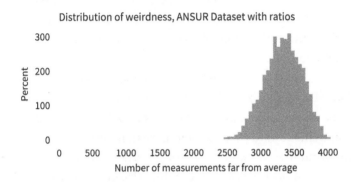

Distribution of weirdness, ANSUR Dataset with ratios

Now, you might notice that the distribution of this weirdness score has the characteristic shape of a Gaussian curve, and that is not a coincidence. Mathematically, as the number of measurements increases, the distribution of weirdness converges to a Gaussian distribution and the width of the distribution gets narrower. In the limit, if we consider the nearly infinite ways people vary, we find that we are all equally weird.

BUT SOME ARE MORE EQUAL THAN OTHERS

In some ways, this chapter is a feel-good story. If at times in your life you have felt different, you are not alone; everyone is different in many ways. And if you thought everyone else was normal, you were wrong—they are all just as weird. But the real world is not like the mathematical world, and I want to be careful not to push this point too far. The idea that we are all equally weird is only true if the different ways of being different are treated the same. And that's not the case.

Considering the measurements in this chapter, some are more visible than others. If your menton-sellion length is in the 90th percentile, people will notice your long face, and if you are a famous actress, Seth MacFarlane will make fun of it. But if your lateral malleolus height is in the 90th percentile, you were probably not teased as a child because your ankle bone is unusually far from the ground.

Among personality traits, some kinds of variation are met more favorably than others. For the most part, we accept introversion and extroversion as equally valid ways to be, but some of the other traits are more laden with value judgments. In general, we consider it better to be conscientious than unreliable and better to be agreeable than obnoxious.

Also, the world is designed to handle some kinds of variation better than others. For example, while introversion might not be considered a moral failing, many introverts find that school and work environments reward extroversion more than might be merited.

The world handles some kinds of physical variation better than others, too. In *Invisible Women*, Caroline Criado Perez writes about many ways our built environment, designed primarily with men in mind, does not always fit women. Safety features in cars, tested on crash dummies with male proportions, protect women less effectively. Many smartphones are difficult to use one-handed if your hands are smaller than average. And, as more than half of the population discovered during the COVID-19 pandemic, a lot of personal protective equipment (PPE) is "based on the sizes and characteristics of male populations from certain countries in Europe and the United States. As a result, most women, and also many men, experience problems finding suitable and comfortable PPE," according to the Trades Union Congress in the United Kingdom.

Furthermore, the way I defined a "weirdness score" is too blunt. To be consistent with Daniels's analysis, I consider only two categories: approximately average or not. In reality, our ability to handle variation has many levels. If you are a little taller or shorter than average, you have to adjust your car seat, but you might not even notice the inconvenience. But if you are too short to stand behind a "standard" lectern or too tall to walk through a door without ducking, you will notice that the world is not designed for you. And at greater extremes, when the mismatch between you and the world makes common tasks difficult or impossible, we call it a disability—and implicitly put the blame on the person, not the built environment.

So, when I say we are all equally weird, my intent is to point out the ways variability defies our intuitive concepts of "normal" and "weird." I don't mean to say that the world treats us as equally weird; it doesn't.

SOURCES AND RELATED READING

- The ANSUR-II dataset is available from the Open Design Lab at Penn State [8]. The measurements are described in "Measurer's Handbook: US Army and Marine Corps Anthropometric Surveys, 2010–2011" [54], which is more interesting than you might expect, especially if you need to know the proper term of address for a Marine Master Gunnery Sergeant.
- My example from a pediatric growth chart is from the Centers for Disease Control and Prevention (CDC) [31].
- You can take a version of the Big Five Personality Test, called the International Personality Item Pool (IPIP), online [13].
- Daniels's paper is available from the Defense Technical Information Center [26].
- Todd Rose wrote about the myth of the average person in *The End of Average* [105].
- Caroline Criado Perez described some of the ways the world is ill-designed for women in an article [95] based on her book *Invisible Women* [94].
- The report on PPE and women is available from the Trades Union Congress [96].
- Sara Hendren wrote about the ways the world is ill-designed for people who are far from average in *What Can a Body Do?* [51].

CHAPTER 2

RELAY RACES AND REVOLVING DOORS

When you run 209 miles, you have a lot of time to think. In 2010, I was a member of a 12-person team that ran a 209-mile relay race in New Hampshire. Long-distance relay races are an unusual format, and this was the first (and last!) time I participated in one. I ran the third leg, so when I joined the race, it had been going for a few hours and runners were spread out over several miles of the course.

After I ran a few miles, I noticed something unusual:

- There were more fast runners in the race than I expected. Several times I was overtaken by runners much faster than me.
- There were also more slow runners than I expected. When I passed other runners, I was often much faster than them.

At first I thought this pattern might reflect the kind of people who sign up for a 209-mile race. Maybe, for some reason, this format appeals primarily to runners who are much faster than average or much slower, and not as much to middle-of-the-pack runners like me.

After the race, with more oxygen available to my brain, I realized that this explanation is wrong. To my embarrassment, I was fooled by a common statistical error, one that I teach students in my classes! The error is called length-biased sampling, and its effect is called the inspection paradox. If you have not heard of it, this chapter will change your life, because once you learn about the inspection paradox, you see it everywhere.

To explain it, I'll start with simple examples, and we will work our way up. Some of the examples are fun, but some are more serious. For example, length-biased sampling shows up in the criminal justice system and distorts our perception of prison sentences and the risk of repeat offenders.

But it's not all bad news; if you are aware of the inspection paradox, sometimes you can use it to measure indirectly quantities that would be hard or impossible to measure directly. As an example, I'll explain a clever system used during the COVID pandemic to track infections and identify superspreaders. But let's start with class sizes.

CLASS SIZE

Suppose you ask college students how big their classes are and average the responses. The result might be 90. But if you ask the college for the average class size, they might say 35. It sounds like someone is lying, but they could both be right.

When you survey students, you oversample large classes; that is, large classes are more likely to appear in your sample than small classes. For example, if there are 10 students in a class, you have 10 chances to sample that class; if there are 100 students, you have 100 chances. In general, if the class size is x, it will be overrepresented in the sample by a factor of x.

That's not necessarily a mistake. If you want to quantify student experience, the average across students might be a more meaningful statistic than the average across classes. But you have to be clear about what you are measuring and how you report it.

The numbers in this example are real. They come from data reported by Purdue University for undergraduate class sizes in the 2013–14 academic year. From their report, I estimated the actual distribution of class sizes. In the following figure, the solid line shows the result as it would be reported by the dean. Most classes are smaller than 50, fewer are between 50 and 100, and only a few are bigger than 100.

Distribution of class sizes

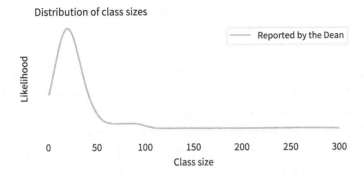

The upper bound in this figure, 300, is just my guess. The original data indicates how many classes are bigger than 100, but it doesn't say how much bigger. For this example, though, we don't have to be too precise.

Now suppose we sample students and ask how big their classes are. We can simulate this process by drawing a random sample from the actual distribution, where the probability of sampling each class is proportional to its size. In the following figure, the dashed line shows the distribution of this simulated sample; the solid line shows the original distribution again.

Distribution of class sizes

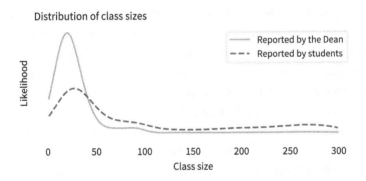

The student sample is less likely to contain classes smaller than 40 and more likely to contain larger classes. And the means of the distributions are substantially different. The mean of the distribution reported by the dean is about 35. The mean of the distribution seen by students is about 90.

A discrepancy like this is called an "inspection paradox" because the data you get depends on how you do the inspection—that is, how you select the sample. If you visit a college and choose a classroom at random (without regard for its size) the class you find inside will have 35 students, on average. But if you choose a student at random and follow them to one of their classes, it will have 90 students, on average. The apparent contradiction is paradoxical in the sense that it is unexpected. It's not a true paradox in the sense used by philosophers, like a self-contradictory statement. But it can certainly be confusing and, if you are not familiar with the phenomenon, hard to explain.

UNBIASING THE DATA

The inspection paradox can be a source of error if you think you measured one thing and accidentally measured another. But in the right hands, it can also be a tool. For example, suppose you are a student at a large university, and all your classes have more than 100 students. The university web page says the average class size is 35, but you suspect that's not accurate. You ask the university for more data, but they won't give it to you. So you decide to collect your own data.

Now, let's assume that you find a good-quality random sample of undergraduates at your university, with equal representation of all departments and grade levels. And let's assume that the students in your sample report the sizes of their classes accurately.

Because you have read this chapter, you know that the sample you collected is biased; that is, it is more likely to include large classes and less likely to include small classes. So you can't just compute the mean of the biased data. That would be the answer to a different question. However, we don't just know that the sample you collected is biased; we know precisely *how* it is biased, and that means we can reverse the process and estimate the unbiased distribution.

In the previous section, I started with the unbiased distribution, as reported by the university, and simulated what would happen if I asked students about their classes. Now I will go the other way, starting with the biased distribution as reported by students and simulating what happens if we sample classes instead. Specifically, I will draw a random sample from the student-reported classes where the probability of choosing each class is *inversely* proportional to its size. The size of the sample I generated is 500, which seems like an amount of data it would be feasible to collect.

The following figure shows the result. Again, the solid line shows the distribution as reported by the dean, and the dashed line shows the biased distribution reported by students. The new, dotted line shows my estimate of the actual distribution, constructed by drawing a sample from the biased distribution.

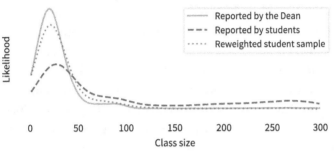

If the estimate were perfect, the solid and dotted lines would be identical. But with a limited sample size, we underestimate the number of small classes by a little and overestimate the number of classes with 50–80 students. Nevertheless, it works pretty well.

This strategy works in other cases where the actual distribution is not available, deliberately or not. If we can collect a good-quality sample from the biased distribution, we can approximate the actual distribution by drawing a sample from the biased data. This process is an example of weighted resampling. It's "weighted" in the sense that some items are given more weight than others—that is, more probability of being sampled. And it's called "resampling" because we're drawing a random sample from something that is itself a random sample.

WHERE'S MY TRAIN?

Another version of the inspection paradox happens when you are waiting for public transportation. Buses and trains are supposed to arrive at regular intervals, but in practice some intervals are longer than others.

You might think, with your luck, that you are more likely to arrive during a long interval. And you're right: a random arrival is more likely to fall in a long interval because, well, it's longer. To quantify this effect, I collected data from the Red Line, which is a subway line in Boston, Massachusetts. The MBTA, which operates the Red Line, provides a real-time data service, which I used to record the arrival times for 70 trains between 4:00 p.m. and 5:00 p.m. over several days.

The shortest gap between trains was less than three minutes; the longest was more than 15. In the following figure, the solid line shows the distribution of the intervals I recorded. This is the actual distribution, in the sense that it is what you would see if you stood on the platform all day and watched the trains go by. It resembles a bell curve, but it is a little pointier near eight minutes; that is, values near the middle are more likely than we would expect from a Gaussian distribution, and values at the extremes are a little less likely. There is actually a term for "pointier than a Gaussian distribution," and it is one of my favorite words: *leptokurtotic.*

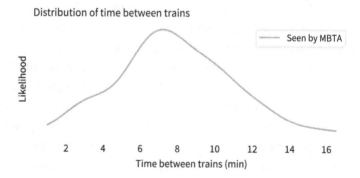

However, and more to my point, this is not the distribution you would see as a passenger. Assuming that you arrive at a random time, without any knowledge about when the next train will arrive, you are more likely to arrive during a long interval; specifically, if an interval is x minutes long, you will oversample it by a factor of x.

As in the previous example with class sizes, I simulated this process, drawing a random sample from the observed intervals with the probability of choosing each interval proportional to its duration. In the following figure, the dashed line shows the result. In this biased distribution, long intervals are relatively more likely, and short intervals are relatively less likely; the effect is that the biased distribution is shifted to the right.

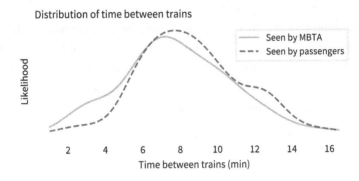

In the actual distribution, the average time between trains is 7.8 minutes; in the biased distribution, as seen by a random passenger, it is 9.2 minutes, about 20% longer. In this example, the discrepancy is not very big; if you ride the train once or twice a day, you might not notice it.

The size of the discrepancy depends on how much variation there is in the distribution. If the intervals between trains were all the same size, there would be no discrepancy at all. But as we'll see in the next example, when the distribution is highly variable, bias in the sampling process can make a big difference.

ARE YOU POPULAR? HINT: NO.

In 1991, the sociologist Scott Feld published a seminal paper on the "friendship paradox," which is the observation that most people have fewer friends than their friends have. He studied real-life social networks, but the same effect appears in online networks. For example, if you choose a random Facebook user and then choose one of their friends at random, the chance is about 80% that the friend has more friends.

The friendship paradox is a form of the inspection paradox. When you choose a random user, every user is equally likely. But when you choose one of their friends, you are more likely to choose someone with a lot of friends. To see why, suppose we visit every person in a social network and ask them to list their friends. If we put all of the lists together, someone with one friend will appear only once, someone with 10 friends will appear 10 times, and in general someone with x friends will appear x times.

To demonstrate the effect, I used data from a sample of about 4000 Facebook users. First I computed the number of friends each user has; in the following figure, the solid line shows this distribution. Most of the users in this sample have fewer than 50 friends, but a few of them have more than 200.

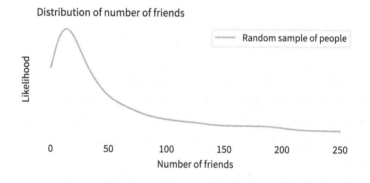

Distribution of number of friends

Then I drew a sample from the list of friends and counted how many friends *they* had. In the following figure, the dashed line shows this distribution. Compared to the unbiased distribution, people with more than 50 friends are overrepresented; people with fewer friends are underrepresented.

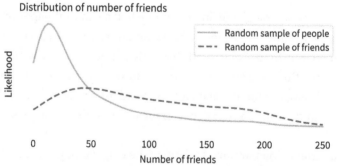

Distribution of number of friends

The difference between the distributions is substantial: in the unbiased sample, the average user has 44 friends; in the biased sample, the average friend has 104, more than twice as many. And if you are a random person in this sample, the probability that your friend is more popular than you is about 76%.

But don't feel too bad; even the most charismatic megafauna have the same problem. If you are a bottlenose dolphin in Doubtful Sound, New Zealand, you interact regularly with between one and 12 other dolphins, according to the researchers who mapped your social network. If I choose one of your "friends" at random and compare them to you, the probability is about 65% that your friend has more friends than you.

You might have noticed that the length in "length-biased sampling" is not always an extent in space. In the class-size example, the "length" of a class is the number of students; in the Red Line example, it's the duration of the interval between trains; in the friendship paradox, it's the number of friends. In the next example, it's the number of people infected by a superspreader.

FINDING SUPERSPREADERS

Fortunately, the inspection paradox can do more than make us feel bad. It can help us find people who feel bad; more specifically, it can help us find people spreading a disease. During the COVID-19 pandemic, you probably heard about the effective reproduction number, R, which is the average number of people infected by each infected person. R is important because it determines the large-scale course of the epidemic. As long as R is greater than one, the number of cases grows exponentially; if it drops below one, the number of cases dwindles toward zero.

However, R is an average, and the average is not the whole story. With COVID-19, as with many other epidemics, there is a lot of variation around the average. According to a news feature in *Nature*, "One study in Hong Kong found that 19% of cases of COVID-19 were responsible for 80% of transmission, and 69% of cases didn't transmit the virus to anyone." In other words, most infections are caused by a small number of superspreaders.

This observation suggests a strategy for contact tracing. When an infected patient is discovered, it is common practice to identify people they have been in contact with who might also be infected. "Forward tracing" is intended to find people the patient might have infected; "backward tracing" is intended to find the person who infected the patient.

Now suppose you are a public health officer trying to slow or stop the spread of a communicable disease. Assuming that you have limited resources to trace contacts and test for the disease—and that's a pretty good assumption—which do you think would be more effective, forward or backward tracing? The inspection paradox suggests that backward tracing is more likely to discover a superspreader and the people they have infected. According to the *Nature* article, "Backward tracing is extremely effective for the coronavirus because of its propensity to be passed on in superspreading events. [. . .] Any new case is more likely to have emerged from a cluster of infections than from one individual, so there's value in going backwards to find out who else was linked to that cluster."

To quantify this effect, let's suppose that 70% of infected people don't infect anyone else, as in the Hong Kong study, and the other 30% infect between one and 15 other people, uniformly distributed. The average of this distribution is 2.4, which is a plausible value of R.

Now suppose we discover an infected patient, trace forward, and find someone the patient infected. On average, we expect this person to infect 2.4 other people. But if we trace backward and find the person who infected the patient, we are more likely to find someone who has infected a lot of people and less likely to find someone who has infected only a few. In fact, the probability that we find any particular spreader is proportional to the number of people they have infected.

By simulating this sampling process, we can compute the distribution we would see by backward tracing. The average of this biased distribution is 10.1, more than four times the average of the unbiased distribution. This result suggests that backward tracing can discover four times more cases than forward tracing, given the same resources.

This example is not just theoretical; Japan adopted this strategy in February 2020. As Michael Lewis describes in *The Premonition*,

> When the Japanese health authorities found a new case, they did not waste their energy asking the infected person for a list of contacts over the previous few days, to determine whom the person might have infected in turn. [. . .] Instead, they asked for a list of people with whom the infected person had interacted further back in time. Find the person who had infected the newly infected person and you might have found a superspreader. Find a superspreader and you could track down the next superspreader before [they] really got going.

So the inspection paradox is not always a nuisance; sometimes we can use it to our advantage.

ROAD RAGE

Now let's get back to the example at the beginning of the chapter, based on my experience running a 209-mile relay race. To recap, I noticed something unusual when I joined the race: when I overtook slower runners, they were usually much slower; and when faster runners passed me, they were usually much faster. At first I thought the distribution of runners had two modes, with many slow runners, many fast runners, and few runners like me in the middle. Then I realized I was being fooled by the inspection paradox.

In long relay races, runners at different speeds end up spread out over the course with no relationship between their speed and location. So if you jump into the middle of the race, the people near you are something like a random sample of the runners in the race.

Considering first the runners behind you, someone running much faster than you is more likely to overtake you during your time on the course than someone just a little bit faster. And, considering the runners in front of you, someone running much slower than you is more likely to be overtaken than someone just a little bit slower. Finally, if someone is running at the same speed as you, you might see them if they happen to be nearby when you join the race, but otherwise there is no chance that you will pass them or them you.

So the sample of runners you see depends on your speed. Specifically, your chance of seeing another runner is proportional to the difference between your speed and theirs. We can simulate this effect using data from a conventional road race. In the following figure, the solid line shows the actual distribution of speeds from the James Joyce Ramble, a 10K race in Massachusetts. This is the distribution a spectator would see if they watched all of the runners go by. The dashed line shows the biased distribution that would be seen by a runner going 11 kilometers per hour (kph).

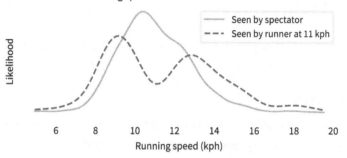

In the actual distribution, there are a lot of runners near 11 kph, but if you run at that speed, you are unlikely to see them. As a result, the biased distribution has few runners near 11 kph and more at the extremes. And it has two modes, one near 9 kph and one near 13 kph. So that explains my oxygen-deprived confusion.

If you are not a runner, you might have noticed the same effect on the highway. You are more likely to see drivers who go too fast or too slow and less likely to see safe, reasonable drivers like yourself. George Carlin summed it up:

> Have you ever noticed when you're driving that anyone driving slower than you is an idiot and anyone driving faster than you is a maniac? [. . .] Because there's certainly no one driving at my speed.

JUST VISITING

Another example of the inspection paradox occurred to me when I read *Orange Is the New Black*, a memoir by Piper Kerman, who spent

13 months in a federal prison. Kerman expresses surprise at the length of the sentences her fellow prisoners are serving. She is right to be surprised, but it turns out that she is not just the victim of an inhumane prison system; she has also been misled by the inspection paradox. If you visit a prison at a random time and choose a random prisoner, you are more likely to find a prisoner with a long sentence. By now, you probably see the pattern: a prisoner with sentence x is overrepresented by a factor of x.

To see what difference it makes, I downloaded data from the US Federal Bureau of Prisons (BOP). Each month, they report the distribution of sentences for current inmates in federal prisons. So the result is a biased sample. However, we can use the reported data to estimate the unbiased distribution, as we did with students and class sizes.

In the following figure, the dashed line shows the distribution of sentences as reported by the BOP. The solid line shows the unbiased distribution I estimated. In the BOP sample, sentences less than three years are underrepresented and longer sentences are over-represented.

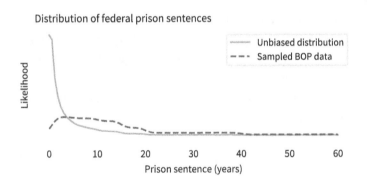

Distribution of federal prison sentences

If you work at a prison, and your job is to check in prisoners on their first day, you see an unbiased sample. If you visit a prison and choose a prisoner at random, you see a biased sample. But what happens if you observe a prison over an interval like 13 months? If the length of your stay is y, the chance of overlapping with a prisoner whose sentence is x is proportional to $x + y$. In the following figure, the dotted line shows what the resulting sample looks like when y is 13 months.

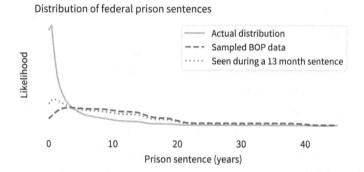

Distribution of federal prison sentences

Someone like Kerman, who served a relatively short sentence, is more likely to see other short-timers, compared to a one-time visitor, and a little less likely to oversample people with long sentences—but not by much. The distribution she observed is substantially different from the actual distribution. We can summarize the differences like this:

- The mean of the actual distribution is 3.6 years; the mean of the biased distribution is almost 13 years, more than three times longer! To a 13-month observer, the mean is about 10 years, still much greater than the actual mean.
- In the actual distribution, about 45% of prisoners have sentences less than a year. If you visit a prison once, fewer than 5% of the prisoners you see are short-timers. If you stay for 13 months, your estimate is better but still not accurate: about 15% of the prisoners you meet are short-timers.

But that's not the only way the inspection paradox distorts our perception of the criminal justice system.

RECIDIVISM

A 2016 paper in the journal *Crime & Delinquency* showed how the inspection paradox affects our estimates of recidivism, that is, the number of people released from prison who later return to prison.

By that definition, prior reports estimated that 45–50% of people released from state prisons returned to prison within three years.

But those reports use "event-based" samples; that is, prisoners selected on the basis of an event like their release from prison. That sampling process is biased because someone imprisoned more than once is more likely to appear in the sample. The alternative is an individual-based sample, where every prisoner is equally likely to appear, regardless of how many sentences they serve.

In an event-based sample, recidivists are oversampled, so the recidivism rate is higher. Using data from the paper, we can find out how much higher. Based on reports from 17 states, collected between 2000 and 2012, the authors compute the number of admissions to prison in an event-based sample, shown in the following figure.

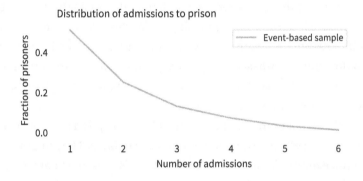

In this sample, 51% of the prisoners served only one sentence; the other 49% were recidivists. So that's consistent with results from previous reports. We can use this data to simulate an individual-based sample. In the following figure, the dashed line shows the result.

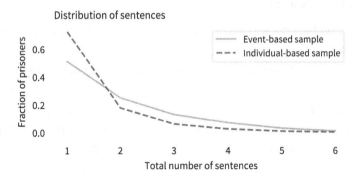

In the individual-based sample, most prisoners serve one sentence; only 28% are recidivists. That's substantially lower than the recidivism rate in the event-based sample, which is 49%. Neither of these statistics is wrong, but they answer different questions.

The individual-based sample tells us what fraction of the people who ever serve time are recidivists. If we want to know whether prison is effective at deterring crime or if we are evaluating a new program intended to reduce recidivism, this might be a useful statistic to consider. By this standard, the outcome is relatively good; as the authors of the paper observe, "most offenders who enter and exit prison do not return."

The event-based sample tells us what fraction of the people released from prison during a short interval are recidivists. This statistic pertains to the risk of recidivism, but might not be the most useful because it does not distinguish people who have already served multiple sentences from those who will. And, due to the inspection paradox, the event-based sample "exaggerates the failure rate of offenders," which "may fuel a pessimistic view that nothing works, or at least that nothing works very well."

A related phenomenon might affect the perception of a police officer who arrests the same person more than once. They might feel like the exercise is futile, but their sample is biased. On any given day, they are more likely to interact with a career criminal, less likely to encounter a one-timer, and even less likely to interact with a law-abiding citizen. If they are not aware of this bias, it might make their perception of criminal justice bleaker than the reality.

IT'S EVERYWHERE

Once you become aware of the inspection paradox, you see it everywhere.

When I teach my Data Science class, I ask students how big their families are and compute the average. Every semester, the result is higher than the natural average. Do my students come from unusually big families? No, my sampling process is biased. If a family has many children, it is more likely that one of them will be in my class. In general, families with x children are overrepresented by a factor of x.

When you call customer service, why is the call center *always* experiencing "higher than normal call volume"? Because when they are busy, many people hear this message; when they are less busy, there are fewer customers to enjoy it.

Airlines complain that too many flights are nearly empty; at the same time, passengers complain that too many planes are full. Both could be true. When a flight is nearly empty, only a few passengers enjoy the extra space. But when a flight is full, many passengers feel the crunch.

In summary, the inspection paradox appears in many domains, sometimes in subtle ways. If you are not aware of it, it can cause statistical errors and lead to invalid conclusions. But in many cases it can be avoided, or even used deliberately as part of an experimental design.

SOURCES AND RELATED READING

- The class-size data from Purdue University is no longer on their web page, but it is available from the Internet Archive [32].
- Information about the Red Line data is available from the MBTA [75].
- Feld's original paper about the friendship paradox is "Why Your Friends Have More Friends Than You Do" [41]). In 2011, a group of researchers published a study of 721 million Facebook users that quantifies the friendship paradox [127]. I don't have access to their data, so I worked with a much smaller subset from a 2021 paper [76], which is available from the Network Data Repository [107].
- John Allen Paulos presented the friendship paradox in *Scientific American* [92]. Steven Strogatz wrote about it, as well as the class-size paradox, in the *New York Times* [117].
- The dolphin data is originally from a paper in *Behavioral Ecology and Sociobiology* [71]; I downloaded it from the Network Data Repository [107].
- Michael Lewis wrote about contact tracing in *The Premonition* [67], based in part on an article in *Nature* [66].
- Zeynep Tufekci wrote about superspreaders in the *Atlantic* [125].

- The running-speed data is originally from Cool Running, now available from the Internet Archive [2].
- I got data on the duration of prison sentences from the website of the Federal Bureau of Prisons [111].
- The paper on recidivism is Rhodes et al., "Following Incarceration, Most Released Offenders Never Return to Prison" [100].

CHAPTER 3

DEFY TRADITION, SAVE THE WORLD

Suppose you are the ruler of a small country where the population is growing quickly. Your advisers warn you that unless this growth slows down, the population will exceed the capacity of the farms and the peasants will starve. The Royal Demographer informs you that the average family size is currently 3; that is, each woman in the kingdom bears three children, on average. He explains that the replacement level is close to 2, so if family sizes decrease by one, the population will level off at a sustainable size.

One of your advisers asks, "What if we make a new law that says every woman has to have fewer children than her mother?" It sounds promising. As a benevolent despot, you are reluctant to restrict your subjects' reproductive freedom, but it seems like such a policy could be effective at reducing family size with minimal imposition.

"Make it so," you say.

Twenty-five years later, you summon the Royal Demographer to find out how things are going. "Your Highness," they say, "I have good news and bad news. The good news is that adherence to the new law has been perfect. Since it was put into effect, every woman in the kingdom has had fewer children than her mother."

"That's amazing," you say. "What's the bad news?"

"The bad news is that the average family size has increased from 3.0 to 3.3, so the population is growing faster than before, and we are running out of food."

"How is that possible?" you ask. "If every woman has fewer children than her mother, family sizes have to get smaller, and population growth has to slow down."

Actually, that's not true. In 1976, Samuel Preston, a demographer at the University of Washington, published a paper that presents three surprising phenomena. The first is the relationship between two measurements of family size: what we get if we ask women how many children they have, and what we get if we ask people how many children their mother had. In general, the average family size is smaller if ask women about their children and larger if we ask children about their mothers. In this chapter, we'll see why and by how much.

The second phenomenon is related to changes in American family sizes during the twentieth century. Suppose you compare family sizes during the Great Depression (roughly the 1930s) and the Baby Boom (1946–64). If you survey women who bore children during these periods, the average family was bigger during the Baby Boom. But if you ask their children how big their families were, the average was bigger during the Depression. In this chapter, we'll see how that happened.

The third phenomenon is what we learned in your imaginary kingdom, which I will call Preston's paradox: Even if every woman has fewer children than her mother, family sizes can get bigger, on average. To see how that's possible, we'll start with the two measurements of family size.

FAMILY SIZE

Families are complicated. A woman might not raise her biological children, and she might raise children she did not bear. Children might not be raised with their biological siblings, and they might be raised with children they are not related to. So the definition of "family size" depends on who's asked and what they're asked.

In the context of population growth, we are often interested in lifetime fertility, so Preston defines "family size" to mean "the total number of children ever born to a woman who has completed childbearing." He suggests that this measure "might more appropriately be termed 'brood size,'" but I will stick with the less zoological term.

In the context of sociology, we are often interested in the size of the "family of orientation," which is the family a person grows up in. If we can't measure families of orientation directly, we can estimate their sizes, at least approximately, if we make the simplifying assumption that most women raise the children they bear. To show how that works, I'll use data from the US Census Bureau.

Every other year, as part of the Current Population Survey (CPS), the Census Bureau surveys a representative sample of women in the United States and asks, among other things, how many children they have ever born. To measure completed family sizes, they select women aged 40–44 (of course, some women bear children in their forties, so these estimates might be a little low). I used their data from 2018 to estimate the current distribution of family sizes. The following figure shows the result. The circles show the percentage of women who bore each number of children. For example, about 15% of the women in the sample had no children; almost 35% of them had two.

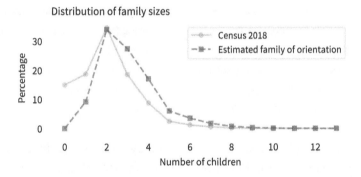

Now, what would we see if we surveyed these children and asked them how many children their mothers had? Large families would be oversampled, small families would be undersampled, and families with no children would not appear at all. In general, a family with k children would appear in the sample k times.

So, if we take the distribution of family size from a sample of women, multiply each bar by the corresponding value of k, and then divide through by the total, the result is the distribution of family size from a sample of children. In the previous figure, the square markers show this distribution. As expected, children would report more large families (three or more children), fewer families with one child, and no families with zero children.

In general, the average family size is smaller if we survey women than if we survey their children. In this example, the average reported by women is close to two; the average reported by their children would be close to three. So you might expect that, if family sizes reported by women get bigger, family sizes reported by children would get bigger, too. But that is not always the case, as we'll see in the next section.

THE DEPRESSION AND THE BABY BOOM

Preston used data from the US Census Bureau to compare family sizes at several points between 1890 and 1970. Among the results, he finds:

- In 1950, the average family size reported by the women in the survey was 2.3. In 1970, it was 2.7. This increase was not surprising, because the first group did most of their childbearing during the Great Depression; the second group, mostly during the Baby Boom.
- However, if we survey the children of these women and ask about their mothers, the average in the 1950 sample would be 4.9; in the 1970 sample it would be 4.5. This decrease was surprising.

According to the women in the survey, families got bigger between 1950 and 1970, by almost half a child. According to their children, families got smaller during the same interval, by almost half a child. It doesn't seem like both can be true, but they are. To understand why, we have to look at the whole distribution of family size, not just the averages.

Preston's paper presents the distributions in a figure, so I used a graph-digitizing tool to extract them numerically. Because this process introduces small errors, I adjusted the results to match the average family sizes Preston computed. The following figure shows these distributions.

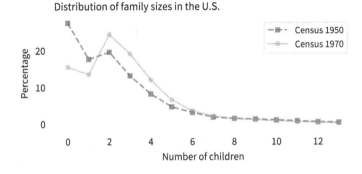

Distribution of family sizes in the U.S.

The biggest differences are in the left part of the curve. The women surveyed in 1950 were substantially more likely to have zero children or one child, compared to women surveyed in 1970. They were less likely to have two to five children and slightly more likely to have nine or more.

Looking at it another way, family size was more variable in the 1950 cohort. The standard deviation of the distribution was about 2.4; in the 1970 cohort, it was about 2.2. This variability is the reason for the difference between the two ways of measuring family size.

As Preston derived, there is a mathematical relationship between the average family size as reported by women, which he calls X, and the average family size as reported by their children, which he calls C:

$$C = X + V/X,$$

where V is the variance of the distribution (the square of standard deviation). Between 1950 and 1970, X got bigger, but V got smaller; and as it turns out, C got smaller, too. That's how family sizes can be bigger if we survey mothers and smaller if we survey their children.

MORE RECENTLY

With more recent census data, we can see what has happened to family sizes since 1970. The following figure shows the average number of children born to women aged 40–44 when they were surveyed. The first cohort was interviewed in 1976, the last in 2018.

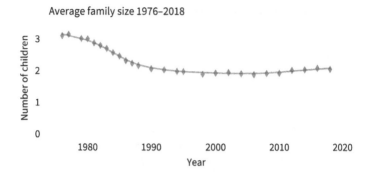

Between 1976 and 1990, average family size fell from more than three to less than two. Since 2010, it has increased a little. To see where those changes came from, let's see how each part of the distribution has changed. The following figure shows the percentage of women with zero, one, or two children, plotted over time.

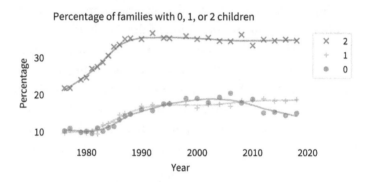

The fraction of small families has increased; most notably, between 1976 and 1990, the percentage of women with two children increased from 22% to 35%. The percentage of women with one child or zero children also increased. The following figure shows how the percentage of large families changed over the same period.

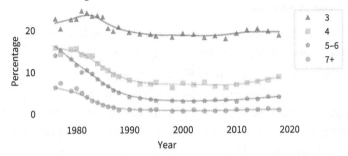

Percentage of families with 3 or more children

The proportion of large families declined substantially, especially the percentage of families with five or six children, which was 14% in 1976 and 4% in 2018. Over the same period, the percentage of families with seven or more children declined from 6% to less than 1%. We can use this data to see how the two measures of family size have changed over time. The following figure shows X, which is the average number of children ever born to the women who were surveyed, and C, which is the average we would measure if we surveyed their children.

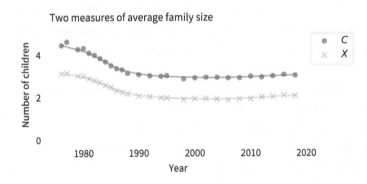

Two measures of average family size

Between 1976 and 1990, the average family size reported by women, X, decreased from 3.1 to 2.0. Over the same period, the average family size reported by children, C, decreased from 4.4 to 2.8. Since 1990, X has been close to two and C has been close to three.

PRESTON'S PARADOX

Now we are ready to explain the surprising result from the imaginary kingdom at the beginning of the chapter. In the scenario, the average family size was three. Then we enacted a new law requiring every woman to have fewer children than her mother. In perfect compliance with the law, every woman had fewer children than her mother. Nevertheless, 25 years later the average family size was 3.3.

How is that possible?

As it happens, the numbers in the scenario are not contrived; they are based on the actual distribution of family size in the United States in 1979. Among the women who were interviewed that year, the average family size was close to three. Of course, nothing like the "fewer than mother" law was implemented in the United States, and 25 years later the average family size was 1.9, not 3.3. But using the actual distribution as a starting place, we can simulate what would have happened in different scenarios.

First, let's suppose that every woman has the same number of children as her mother. At first, you might think that the distribution of family size would not change, but that's not right. In fact, family sizes would increase quickly. As a small example, suppose there are only two families, one with two children and the other with four, so the average family size is three.

Assuming that half of the children are girls, in the next generation there would be one woman from a two-child family and two children from a four-child family. If each of these women has the same number of children as her mother, one would have two children and the other two would have four. So the average family size would be 3.33.

In the next generation, there would be one woman with two children, again, but there would be four women with four children. So the average family size would be 3.6. In successive generations, the number of four-child families would increase exponentially, and the average family size would quickly approach four.

We can do the same calculation with a more realistic distribution. In fact, it is the same calculation we used to compute the distribution of family size as reported by children. In the same way large families are overrepresented when we survey children, large families are overrepresented in the next generation if every woman has the same number of children as her mother. The following figure shows the actual distribution of family sizes from 1979 (solid line) and the result of this calculation, which simulates the "same as mother" scenario (dashed line).

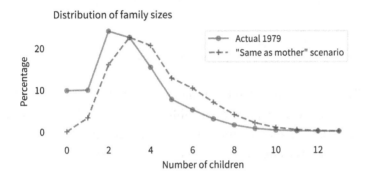

In the next generation, there are fewer small families and more big families, so the distribution is shifted to the right. The mean of the original distribution is 3; the mean of the simulated distribution is 4.3. If we repeated the process, the average family size would be 5.2 in the next generation, 6.1 in the next, and 6.9 in the next. Eventually, nearly all women would have 13 children, which is the maximum in this dataset.

Observing this pattern, Preston explains, "The implications for population growth are obvious. Members of each generation must, on average, bear substantially fewer children than were born into their own family of orientation merely to keep population fertility rates constant." So let's see what happens if we follow Preston's advice.

ONE CHILD FEWER

Suppose every woman has precisely one child fewer than her mother. We can simulate this behavior by computing the biased distribution, as in the previous section, and then shifting the result one child to the left. The following figure shows the actual distribution from 1979 again, along with the result of this simulation.

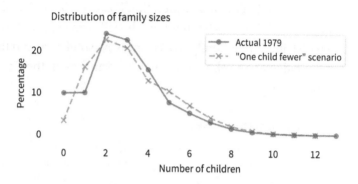

Distribution of family sizes

The simulated distribution has been length-biased, which increases the average, and then shifted to the left, which decreases the average. In general, the net effect could be an increase or a decrease; in this example, it's an increase from 3.0 to 3.3.

IN THE LONG RUN

If we repeat this process and simulate the next generation, the average family size increases again, to 3.5. In the next generation it increases to 3.8. It might seem like it would increase forever, but if every woman has one child fewer than her mother, in each generation the *maximum* family size decreases by one. So eventually the *average* comes down. The following figure shows the average family size over 10 generations, where Generation Zero follows the actual distribution from 1979.

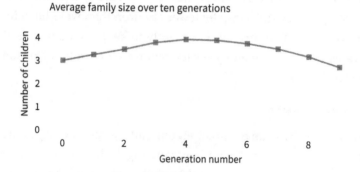

Average family size over ten generations

Family size would increase for four generations, peaking at 3.9. Then it would take another five generations to fall below the starting value. At 25 years per generation, that means it would take more than 200 years for the new law to have the desired effect.

IN REALITY

Of course, no such law was in effect in the United States between 1979 and 1990. Nevertheless, the average family size fell from close to 3 to close to 2. That's an enormous change in less than one generation. How is that even possible? We've seen that it is not enough if every woman has one child fewer than her mother; the actual difference must have been larger.

In fact, the difference was about 2.3. To see why, remember that the average family size, as seen by the children of the women surveyed in 1979, was 4.3. But when those children grew up and the women among them were surveyed in 1990, they reported an average family size close to 2. So on average, they had 2.3 fewer children than their mothers.

As Preston explains, "A major intergenerational change at the individual level is required in order to maintain intergenerational stability at the aggregate level. . . . Those who exhibit the most traditional behavior with respect to marriage and women's roles will always be overrepresented as parents of the next generation, and a perpetual disaffiliation from their model by offspring is required in order to avert an increase in traditionalism for the population as a whole." In other words, children *have to* reject the childbearing example of their mothers; otherwise, we are all doomed.

IN THE PRESENT

The women surveyed in 1990 rejected the childbearing example of their mothers emphatically. On average, each woman had 2.3 fewer children than her mother. If that pattern had continued for another generation, the average family size in 2018 would have been about 0.8. But it wasn't. In fact, the average family size in 2018 was very close to 2, just as in 1990. So how did that happen?

As it turns out, this is close to what we would expect if every woman had one child fewer than her mother. The following figure shows the actual distribution in 2018 compared to the result if we start with the 1990 distribution and simulate the "one child fewer" scenario.

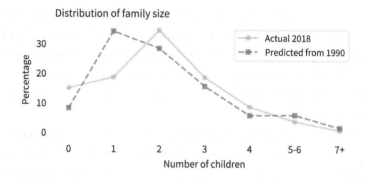

Distribution of family size

The means of the two distributions are almost the same, but the shapes are different. In reality, there were more families with zero and two children in 1990 than the simulation predicts and fewer one-child families. But at least on average, it seems like women in the US have been following the "one child fewer" policy for the past 30 years.

The scenario at the beginning of this chapter is meant to be light-hearted, but in reality governments in many places and times have enacted policies meant to control family sizes and population growth. Most famously, China implemented a one-child policy in 1980 that imposed severe penalties on families with more than one child. Of course, this policy is objectionable to anyone who considers reproductive freedom a fundamental human right. But even as a practical matter, the unintended consequences were profound.

Rather than catalog them, I will mention one that is particularly ironic: while this policy was in effect, economic and social forces reduced the average *desired* family size so much that, when the policy was relaxed in 2015 and again in 2021, average lifetime fertility increased to only 1.3, far below the level needed to keep the population constant, near 2.1. Since then, China has implemented new policies intended to *increase* family sizes, but it is not clear whether they will have much effect. Demographers predict that by the time you read this, the population of China will probably be shrinking. The consequences of the one-child policy are widespread and will affect China and the rest of the world for a long time.

SOURCES AND RELATED READING

- Preston's paper is "Family Sizes of Children and Family Sizes of Women" [99].
- Documentation of the CPS Supplemental surveys is on the website of the Census Bureau [118]. The data I used is "Distribution of Women Age 40–50 by Number of Children Ever Born and Marital Status: CPS, Selected Years, 1976–2018" [1].
- In September 2022, the *Economist* wrote about China's one-child policy and recent efforts to increase family sizes [19].

CHAPTER 4

EXTREMES, OUTLIERS, AND GOATS

In chapter 1 we saw that many measurements in the natural world follow a Gaussian distribution, and I posited an explanation: If we add up many random factors, the distribution of the sum tends to be Gaussian. Gaussian distributions are so common that most people have an intuition for how they behave. Thinking about the distribution of height, for example, we expect to find about the same number of people above and below the mean, and we expect the distribution to extend the same distance in both directions. Also, we don't expect the distribution to extend very far. For example, the average height in the United States is about 170 cm. The tallest person ever was 272 cm, which is certainly tall, but it is only 60% greater than the mean. And in all likelihood, no one will ever be substantially taller.

But not all distributions are Gaussian, and many of them defy this intuition. In this chapter, we will see that adult weights do not follow a Gaussian distribution, but their logarithms do, which means that the distribution is "lognormal." As a result, the heaviest people are much heavier than we would expect in a Gaussian distribution. Unlike in the distribution of height, we find people in the distribution of weight who are twice the mean or more. The heaviest person ever reliably measured was almost eight times the current average in the United States.

Many other measurements from natural and engineered systems also follow lognormal distributions. I'll present some of them, including running speeds and chess ratings. And I'll posit an explanation: If we multiply many random factors together, the distribution of the product tends to be lognormal.

These distributions have implications for the way we think about extreme values and outliers. In a lognormal distribution, the heaviest people are heavier, the fastest runners are faster, and the best chess players are much better than they would be in a Gaussian distribution. As an example, suppose we start with an absolute beginner at chess, named A. You could easily find a more experienced player, named B, who would beat A 90% of the time. And it would not be hard to find a player C who could beat B 90% of the time; in fact, C would be pretty close to average. Then you could find a player D who could beat C 90% of the time and a player E who could beat D about as often.

In a Gaussian distribution, that's about all you would get. If A is a rank beginner, and C is average, E would be one of the best, and it would be hard to find someone substantially better. But the distribution of chess skill is lognormal, and it extends farther to the right than a Gaussian. In fact, we can find a player F who beats E, a player G who beats F, and a player H who beats G, at each step more than 90% of the time. And the world champion in this distribution would still beat H almost 90% of the time.

Later in the chapter we'll see where these numbers come from, and we'll get some insight into outliers and what it takes to be one. But let's start with the distribution of weight.

THE EXCEPTION

In chapter 1, I presented bodily measurements from the Anthropometric Survey of US Army Personnel (ANSUR) and showed that a Gaussian model fits almost all of them. In fact, if you take almost any measurement from almost any species, the distribution tends to be Gaussian.

But there is one exception: weight. To demonstrate the point, I'll use data from the Behavioral Risk Factor Surveillance System (BRFSS), which is an annual survey run by the US Centers for Disease Control and Prevention (CDC). The 2020 dataset includes information about demographics, health, and health risks from a large, representative sample of adults in the United States: a total of 195,055 men and 205,903 women. Among the data are the self-reported weights of the respondents, recorded in kilograms.

The following figure shows the distribution of these weights, represented using cumulative distribution functions (CDFs), which I introduced in chapter 1. The lighter lines show Gaussian models that best fit the data; the crosses at the top show the locations of the largest weights we expect to find in samples this size from a Gaussian distribution.

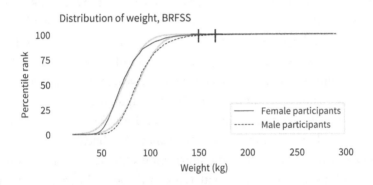

The Gaussian models don't fit these measurements particularly well, and the extremes are more extreme. If these distributions were Gaussian, the heaviest woman in the sample would be about 150 kg, and the heaviest man would be 167 kg. In fact, the heaviest woman is 286 kg and the heaviest man is 290 kg. These discrepancies suggest that the distributions of weight are not Gaussian.

We can get a hint about what's going on from Philip Gingerich, a now-retired professor of Ecology and Evolutionary Biology at the University of Michigan. In 2000, he wrote about two kinds of variation we see in measurements from biological systems. On one hand, we find measurements like the ones in chapter 1 that follow a Gaussian curve, which is symmetric; that is, the tail extends about the same distance to the left and to the right. On the other hand, we find measurements like weight that are skewed; that is, the tail extends farther to the right than to the left. This observation goes back to the earliest history of statistics, and so does a remarkably simple remedy: if we compute the logarithms of the values and plot their distribution, the result is well modeled by—you guessed it—a Gaussian curve.

Depending on your academic and professional background, logarithms might be second nature to you, or they might be something you remember vaguely from high school. For the second group, here's a reminder. If you raise 10 to some power x, and the result is y, that means that x is the logarithm, base 10, of y. For example, 10 to the power of 2 is 100, so 2 is the logarithm, base 10, of 100 (I'll drop the "base 10" from here on, and call a logarithm by its nickname, log). Similarly, 10 to the power of 3 is 1000, so 3 is the log of 1000, and 4 is the log of 10,000. You might notice a pattern: for numbers that are powers of 10, the log is the number of zeros.

For numbers that are not powers of 10, we can still compute logs, but the results are not integers. For example, the logarithm of 50 is 1.7, which is between the logs of 10 and 100, but not equidistant between them. And the log of 150 is about 2.2, between the logs of 100 and 1000. The following figure shows the distributions of the logarithms of weight along with Gaussian models that best fit them.

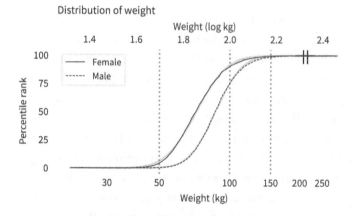

At the top of the figure, the labels indicate the logarithms. At the bottom, the labels indicate the weights themselves. Notice that the logarithms are equally spaced, but the weights are not. The vertical dotted lines show the correspondence between the weights and their logarithms for the three examples I just mentioned: 50 kg, 100 kg, and 150 kg. The Gaussian models fit the distributions so well that the lines representing them are barely visible. Values like this, whose logs follow a Gaussian model, are called "lognormal."

Again, the crosses at the top show the largest values we expect from distributions with these sample sizes. They are a better match for the data than the crosses in the previous figure, although it looks like the heaviest people in this dataset are still heavier than we would expect from the lognormal model.

So, why is the distribution of weight lognormal? There are two possibilities: maybe we're born this way, or maybe we grow into it. In the next section we'll see that the distribution of birth weights follows a Gaussian model, which suggests that we're born Gaussian and grow up to be lognormal. In the following sections I'll try to explain why.

BIRTH WEIGHTS ARE GAUSSIAN

To explore the distribution of birth weight, I'll use data from the National Survey of Family Growth (NSFG), which is run by the CDC, the same people we have to thank for the BRFSS. Between 2015 and 2017, the NSFG collected data from a representative sample of women in the United States. Among other information, the survey records the birth weights of their children. Excluding babies who were born preterm, I selected weights for 3430 male and 3379 female babies. The following figure shows the distributions of these birth weights along with Gaussian curves that best fit them.

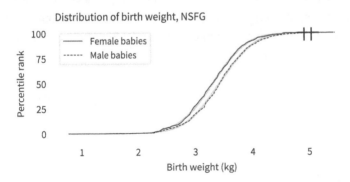

The Gaussian model fits the distribution of birth weights well. The heaviest babies are a little heavier than we would expect from a Gaussian distribution with this sample size, but the distributions are roughly symmetric, not skewed to the right like the distribution of adult weight. So, if we are not born lognormal, that means we grow into it. In the next section, I'll propose a model that might explain how.

SIMULATING WEIGHT GAIN

As a simple model of human growth, suppose people gain several pounds per year from birth until they turn 40. And let's suppose their annual weight gain is a random quantity; that is, some years they gain a lot, some years less. If we add up these random gains, we expect the total to follow a Gaussian distribution because, according to the Central Limit Theorem, when you add up a large number of random values, the sum tends to be Gaussian. So, if the distribution of adult weight were Gaussian, that would explain it.

But we've seen that it is not, so let's try something else: what if annual weight gain is proportional to weight? In other words, instead of gaining a few pounds, what if you gained a few percent of your current weight? For example, if someone who weighs 50 kg gains 1 kg in a year; someone who weighs 100 kg might gain 2 kg. Let's see what effect these proportional gains have on the results.

I wrote a simulation that starts with the birth weights we saw in the previous section and generates 40 random growth rates, one for each year from birth to age 40. The growth rates vary from about 20–40%; in other words, each simulated person gains 20–40% of their current weight each year. This is not a realistic model, but let's see what we get. The following figure shows the results and a Gaussian model, both on a log scale.

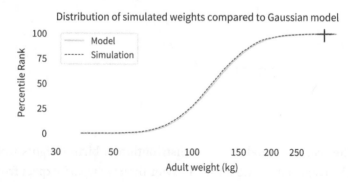

Distribution of simulated weights compared to Gaussian model

The Gaussian model fits the logarithms of the simulated weights, which means that the distribution is lognormal. This outcome is explained by a corollary of the Central Limit Theorem: If you multiply together a large number of random values, the distribution of the product tends to be lognormal.

It might not be obvious that the simulation multiplies random quantities, but that's what happens when we apply successive growth rates. For example, suppose someone weighs 50 kg at the beginning of a simulated year. If they gain 20%, their weight at the end of the year would be 50 kg times 120%, which is 60 kg. If they gain 30% during the next year, their weight at the end would be 60 kg times 130%, which is 78 kg.

In two years, they gained a total of 28 kg, which is 56% of their starting weight. Where did 56% come from? It's not the sum of 20% and 30%; rather, it comes from the product of 120% and 130%, which is 156%. So their weight at the end of the simulation is the product of their starting weight and 40 random growth rates. As a result, the distribution of simulated weights is approximately lognormal. By adjusting the range of growth rates, we can tune the simulation to match the data. The following figure shows the actual distributions from the BRFSS along with the simulation results.

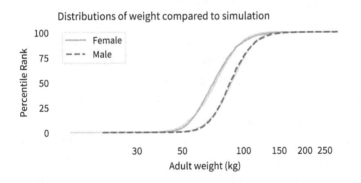

It might not be obvious

The results from the simulation are a good match for the actual distributions. Of course, the simulation is not a realistic model of how people grow from birth to adulthood. We could make the model more realistic by generating higher growth rates for the first 15 years and then lower growth rates thereafter. But these details don't change the shape of the distribution; it's lognormal either way. These simulations demonstrate one of the mechanisms that can produce lognormal distributions: proportional growth. If each person's annual weight gain is proportional to their current weight, the distribution of their weights will tend toward lognormal.

Now, if height is Gaussian and weight is lognormal, what difference does it make? Qualitatively, the biggest difference is that the Gaussian distribution is symmetric, so the tallest people and the shortest are about the same distance from the mean. In the BRFSS dataset, the average height is 170 cm. The 99th percentile is about 23 cm above the mean and the 1st percentile is about 20 cm below, so that's roughly symmetric. Taking it to the extremes, the tallest person ever reliably measured was Robert Wadlow, who was 272 cm, which is 102 cm from the mean. And the shortest adult was Chandra Bahadur Dangi, who was 55 cm, which is 115 cm from the mean. That's pretty close to symmetric, too.

On the other hand, the distribution of weights is not symmetric: the heaviest people are substantially farther from the mean than the lightest. In the United States, the average weight is about 82 kg. The heaviest person in the BRFSS dataset is 64 kg above average, but the lightest is only 36 kg below. Taking it to the extremes, the heaviest person person ever reliably measured was, regrettably, 553 kg heavier than the average. In order to be the same distance below the mean, the weight of the lightest person would be 471 kg lighter than zero, which is impossible. So the Gaussian model of weight is not just a bad fit for the data, it produces absurdities.

In contrast, on a log scale, the distribution of weight is close to symmetric. In the adult weights from the BRFSS, the 99th percentile of the logarithms is 0.26 above the mean and the 1st percentile is 0.24 below the mean. At the extremes, the log of the heaviest weight is 0.9 above the mean. In order for the lightest person to be the same distance below the mean, they would have to weigh 10 kg, which might sound impossible, but the lightest adult, according to Guinness World Records, was 2.1 kg at age 17. I'm not sure how reliable that measurement is, but it is at least close to the minimum we expect based on symmetry of the logarithms.

RUNNING SPEEDS

If you are a fan of the Atlanta Braves, a Major League Baseball team, or if you watch enough videos on the internet, you have probably seen one of the most popular forms of between-inning entertainment: a foot race between one of the fans and a spandex-suit-wearing mascot called the Freeze.

The route of the race is the dirt track that runs across the outfield, a distance of about 160 meters, which the Freeze runs in less than 20 seconds. To keep things interesting, the fan gets a head start of about 5 seconds. That might not seem like a lot, but if you watch one of these races, this lead seems insurmountable. However, when the Freeze starts running, you immediately see the difference between a pretty good runner and a very good runner. With few exceptions, the Freeze runs down the fan, overtakes them, and coasts to the finish line with seconds to spare.

But as fast as he is, the Freeze is not even a professional runner; he is a member of the Braves' ground crew named Nigel Talton. In college, he ran 200 meters in 21.66 seconds, which is very good. But the 200-meter collegiate record is 20.1 seconds, set by Wallace Spearmon in 2005, and the current world record is 19.19 seconds, set by Usain Bolt in 2009.

To put all that in perspective, let's start with me. For a middle-aged man, I am a decent runner. When I was 42 years old, I ran my best-ever 10-kilometer race in 42:44, which was faster than 94% of the other runners who showed up for a local 10K. Around that time, I could run 200 meters in about 30 seconds (with wind assistance). But a good high school runner is faster than me. At a recent meet, the fastest girl at a nearby high school ran 200 meters in about 27 seconds, and the fastest boy ran under 24 seconds.

So, in terms of speed, a fast high school girl is 11% faster than me, and a fast high school boy is 12% faster than her; Nigel Talton, in his prime, was 11% faster than the high school boy, Wallace Spearmon was about 8% faster than Talton, and Usain Bolt is about 5% faster than Spearmon. Unless you are Usain Bolt, there is always someone faster than you, and not just a little bit faster; they are much faster.

The reason, as you might suspect by now, is that the distribution of running speed is not Gaussian. It is more like lognormal.

To demonstrate, I'll use data from the James Joyce Ramble, which is the 10-kilometer race where I ran my previously mentioned personal record time. I downloaded the times for the 1592 finishers and converted them to speeds in kilometers per hour. The following figure shows the distribution of these speeds on a logarithmic scale, along with a Gaussian model I fit to the data.

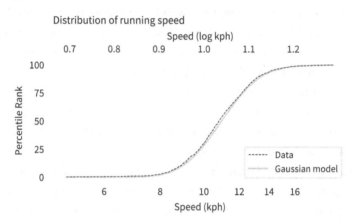

The logarithms follow a Gaussian distribution, which means the speeds themselves are lognormal. You might wonder why. Well, I have a theory, based on the following assumptions:

- First, everyone has a maximum speed they are capable of running, assuming that they train effectively.
- Second, these speed limits can depend on many factors, including height and weight, fast- and slow-twitch muscle mass, cardiovascular conditioning, flexibility and elasticity, and probably more.
- Finally, the way these factors interact tends to be multiplicative; that is, each person's speed limit depends on the product of multiple factors.

Here's why I think speed depends on a product rather than a sum of factors. If all of your factors are good, you are fast; if any of them are bad, you are slow. Mathematically, the operation that has this property is multiplication. For example, suppose there are only two factors, measured on a scale from 0 to 1, and each person's speed limit is determined by their product. Let's consider three hypothetical people:

- The first person scores high on both factors, let's say 0.9. The product of these factors is 0.81, so they would be fast.
- The second person scores relatively low on both factors, let's say 0.3. The product is 0.09, so they would be quite slow.

So far, this is not surprising: if you are good in every way, you are fast; if you are bad in every way, you are slow. But what if you are good in some ways and bad in others?

- The third person scores 0.9 on one factor and 0.3 on the other. The product is 0.27, so that person is a little bit faster than someone who scores low on both factors, but much slower than someone who scores high on both.

That's a property of multiplication: the product depends most strongly on the smallest factor. And as the number of factors increases, the effect becomes more dramatic.

To simulate this mechanism, I generated five random factors from a Gaussian distribution and multiplied them together. I adjusted the mean and standard deviation of the Gaussians so that the resulting distribution fit the data; the following figure shows the results.

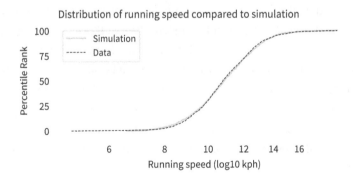

The simulation results fit the data well. So this example demonstrates a second mechanism that can produce lognormal distributions: the limiting power of the weakest link. If there are at least five factors that affect running speed, and each person's limit depends on their worst factor, that would explain why the distribution of running speed is lognormal.

I suspect that distributions of many other skills are also lognormal, for similar reasons. Unfortunately, most abilities are not as easy to measure as running speed, but some are. For example, chess-playing skill can be quantified using the Elo rating system, which we'll explore in the next section.

CHESS RANKINGS

In the Elo chess-rating system, every player is assigned a score that reflects their ability. These scores are updated after every game. If you win, your score goes up; if you lose, it goes down. The size of the increase or decrease depends on your opponent's score. If you beat a player with a higher score, your score might go up a lot; if you beat a player with a lower score, it might barely change. Most scores are in the range from 100 to about 3000, although in theory there is no lower or upper bound.

By themselves, the scores don't mean very much; what matters is the difference in scores between two players, which can be used to compute the probability that one beats the other. For example, if the difference in scores is 400, we expect the higher-rated player to win about 90% of the time.

If the distribution of chess skill is lognormal, and if Elo scores quantify this skill, we expect the distribution of Elo scores to be lognormal. To find out, I collected data from Chess.com, which is a popular internet chess server that hosts individual games and tournaments for players from all over the world. Their leader board shows the distribution of Elo ratings for almost six million players who have used their service. The following figure shows the distribution of these scores on a log scale, along with a lognormal model.

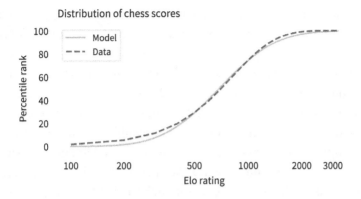

The lognormal model does not fit the data particularly well. But that might be misleading, because unlike running speeds, Elo scores have no natural zero point. The conventional zero point was chosen arbitrarily, which means we can shift it up or down without changing what the scores mean relative to each other. With that in mind, suppose we shift the entire scale so that the lowest point is 550 rather than 100. The following figure shows the distribution of these shifted scores on a log scale, along with a lognormal model. With this adjustment, the lognormal model fits the data well.

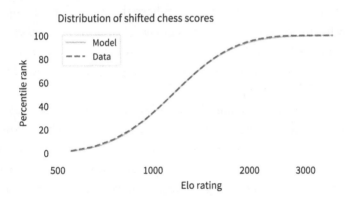

Now we've seen two explanations for lognormal distributions: proportional growth and weakest links. Which one determines the distribution of abilities like chess? I think both mechanisms are plausible. As you get better at chess, you have opportunities to play against better opponents and learn from the experience. You also gain the ability to learn from others; books and articles that are inscrutable to beginners become invaluable to experts. As you understand more, you are able to learn faster, so the growth rate of your skill might be proportional to your current level.

At the same time, lifetime achievement in chess can be limited by many factors. Success requires some combination of natural abilities, opportunity, passion, and discipline. If you are good at all of them, you might become a world-class player. If you lack any of them, you will not. The way these factors interact is like multiplication, where the outcome is most strongly affected by the weakest link.

These mechanisms shape the distribution of ability in other fields, even the ones that are harder to measure, like musical ability. As you gain musical experience, you play with better musicians and work with better teachers. As in chess, you can benefit from more advanced resources. And, as in almost any endeavor, you learn how to learn.

At the same time, there are many factors that can limit musical achievement. One person might have a bad ear or poor dexterity. Another might find that they don't love music enough, or they love something else more. One might not have the resources and opportunity to pursue music; another might lack the discipline and tenacity to stick with it. If you have the necessary aptitude, opportunity, and personal attributes, you could be a world-class musician; if you lack any of them, you probably can't.

If you have read Malcolm Gladwell's book, *Outliers*, this conclusion might be disappointing. Based on examples and research on expert performance, Gladwell suggests that it takes 10,000 hours of effective practice to achieve world-class mastery in almost any field. Referring to a study of violinists led by the psychologist K. Anders Ericsson, Gladwell writes, "The striking thing [. . .] is that he and his colleagues couldn't find any 'naturals,' musicians who floated effortlessly to the top while practicing a fraction of the time their peers did. Nor could they find any 'grinds,' people who worked harder than everyone else, yet just didn't have what it takes to break the top ranks."

The key to success, Gladwell concludes, is many hours of practice. The source of the number 10,000 seems to be neurologist Daniel Levitin, quoted by Gladwell: "In study after study, of composers, basketball players, fiction writers, ice skaters, concert pianists, chess players, master criminals, and what have you, this number comes up again and again. [. . .] No one has yet found a case in which true world-class expertise was accomplished in less time." The core claim of the rule is that 10,000 hours of practice is *necessary* to achieve expertise. Of course, as Ericsson writes in a commentary, "There is nothing magical about exactly 10,000 hours." But it is probably true that no world-class musician has practiced substantially less.

However, some people have taken the rule to mean that 10,000 hours is *sufficient* to achieve expertise. In this interpretation, anyone can master any field; all they have to do is practice! Well, in running and many other athletic areas, that is obviously not true. And I doubt it is true in chess, music, or many other fields. Natural talent is not enough to achieve world-level performance without practice, but that doesn't mean it is irrelevant. For most people in most fields, natural attributes and circumstances impose an upper limit on performance.

In his commentary, Ericsson summarizes research showing the importance of "motivation and the original enjoyment of the activities in the domain and, even more important, [. . .] inevitable differences in the capacity to engage in hard work (deliberate practice)." In other words, the thing that distinguishes a world-class violinist from everyone else is not 10,000 hours of practice, but the passion, opportunity, and discipline it takes to spend 10,000 hours doing anything.

THE GREATEST OF ALL TIME

Lognormal distributions of ability might explain an otherwise surprising phenomenon: in many fields of endeavor, there is one person widely regarded as the "greatest of all time," or GOAT. For example, in hockey, Wayne Gretzky is the GOAT, and it would be hard to find someone who knows hockey and disagrees. In basketball, it's Michael Jordan; in women's tennis, Serena Williams; and so on for most sports. Some cases are more controversial than others, but even when there are a few contenders for the title, there are only a few.

And more often than not, these top performers are not just a little better than the rest; they are a *lot* better. For example, in his career in the National Hockey League, Wayne Gretzky scored 2857 points (the total of goals and assists). The player in second place scored 1921. The magnitude of this difference is surprising, in part, because it is not what we would get from a Gaussian distribution.

To demonstrate this point, I generated a random sample of 100,000 people from a lognormal distribution loosely based on chess ratings. Then I generated a sample from a Gaussian distribution with the same mean and variance. The following figure shows the results.

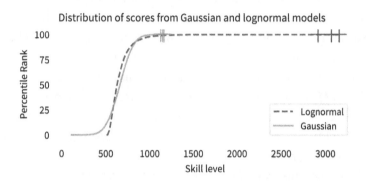

The mean and variance of these distributions are about the same, but the shapes are different: the Gaussian distribution extends a little farther to the left, and the lognormal distribution extends much farther to the right. The crosses indicate the top three scorers in each sample. In the Gaussian distribution, the top three scores are 1123, 1146, and 1161. They are barely distinguishable in the figure, and if we think of them as Elo scores, there is not much difference between them.

According to the Elo formula, we expect the top player to beat the third-ranked player about 55% of the time. In the lognormal distribution, the top three scores are 2913, 3066, and 3155. They are clearly distinct in the figure and substantially different in practice. In this example, we expect the top player to beat number three about 80% of the time.

In reality, the top-rated chess players in the world are more tightly clustered than my simulated players, so this example is not entirely realistic. Even so, Garry Kasparov is widely considered to be the greatest chess player of all time. The current world champion, Magnus Carlsen, might overtake him in another decade, but even he acknowledges that he is not there yet.

Less well known, but more dominant, is Marion Tinsley, who was the checkers (a.k.a. draughts) world champion from 1955 to 1958, withdrew from competition for almost 20 years—partly for lack of competition—and then reigned uninterrupted from 1975 to 1991. Between 1950 and his death in 1995, he lost only seven games, two of them to a computer. The man who programmed the computer thought Tinsley was "an aberration of nature." Marion Tinsley might have been the greatest GOAT of all time, but I'm not sure that makes him an aberration. Rather, he is an example of the natural behavior of lognormal distributions:

- In a lognormal distribution, the outliers are farther from average than in a Gaussian distribution, which is why ordinary runners can't beat the Freeze, even with a head start.
- And the margin between the top performer and the runner-up is wider than it would be in a Gaussian distribution, which is why the greatest of all time is, in many fields, an outlier among outliers.

WHAT SHOULD YOU DO?

If you are not sure what to do with your life, the lognormal distribution can help. Suppose you have the good fortune to be offered three jobs, and you are trying to decide which one to take. One of the companies is working on a problem you think is important, but you are not sure whether they will have much impact on it. The second is also working on an important problem, and you think they will have impact, but you are not sure how long that impact will last. The third company is likely to have long-lasting impact, in your estimation, but they are working on a problem you think is less important. If your goal is to maximize the positive impact of your work, which job should you take?

What I have just outlined is called the significance-persistence-contingency framework, or SPC for short. In this framework, the total future impact of the effort you allocate to a problem is the product of these three factors: significance is the positive effect during each unit of time where the problem is solved, persistence is length of time the solution is applicable, and contingency is the difference between the outcome with or without your effort. If a project scores high on all three of these factors, it can have a large effect on the future; if it scores low on any of them, its impact will be limited.

In an extension of this framework, we can decompose contingency into two additional factors: tractability and neglectedness. If a problem is not tractable, it is not contingent, because it will not get solved regardless of effort. And if it is tractable and not neglected, it is not contingent because it will get solved whether you work on it or not. So there are at least five factors that might limit a project's impact in the world, and there are probably five more that affect your ability to impact the project, including your skills, personal attributes, and circumstances.

In general, the way these factors interact is like multiplication: if any of them are low, the product is low; the only way for the product to be high is if all of the factors are high. As a result, we have reason to think that the distribution of impact, across a population of all the projects you might work on, is lognormal. Most projects have modest potential for impact, but a few can change the world.

So what does that mean for your job search? It implies that—if you want to maximize your impact—you should not take the first job you find, but spend time looking. And it suggests that you should not stay in one job for your entire career; you should continue to look for opportunities and change jobs when you find something better. These are some of the principles that underlie the 80,000 Hours project, which is a collection of online resources intended to help people think about how to best spend the approximately 80,000 hours of their working lives.

SOURCES AND RELATED READING

- The ANSUR-II dataset is available from the Open Design Lab at Penn State [8].

- Data from the Behavioral Risk Factor Surveillance System (BRFSS) is available from the Centers for Disease Control and Prevention (CDC) [11].

- Gingerich's paper comparing Gaussian and lognormal models for anthropometric measurements is "Arithmetic or Geometric Normality of Biological Variation" [45].

- The running-speed data is originally from Cool Running, now available from the Internet Archive [2].

- The chess data is from the Chess.com leaderboard [47].

- The origin of the 10,000-hour claim is Daniel Levitin's book *This Is Your Brain on Music* [65], quoted by Gladwell in *Outliers* [46]. Ericsson's commentary on the topic is "Training History, Deliberate Practice and Elite Sports Performance" [40].

- Oliver Roeder wrote about Marion Tinsley in *Seven Games* [104]. Dr. Jonathan Shaeffer, who wrote the program that won two games against Tinsley, wrote a memorial for him [110].

- The SPC framework is described in a technical report from the Global Priorities Institute [72] and the 80,000 Hours website [124].

CHAPTER 5

BETTER THAN NEW

Suppose you work in a hospital, and one day you have lunch with three of your colleagues. One is a facilities engineer working on a new lighting system, one is an obstetrician who works in the maternity ward, and one is an oncologist who works with cancer patients. While you all enjoy the hospital food, each of them poses a statistical puzzle.

The engineer says they are replacing old incandescent light bulbs with LED bulbs, and they've decided to replace the oldest bulbs first. According to previous tests, the bulbs last 1400 hours on average. So, they ask, which do you think will last longer: a new bulb or one that has already been lit for 1000 hours? Sensing a trick question, you ask if the new bulb might be defective. The engineer says, "No, let's assume we've confirmed that it works." "In that case," you say, "I think the new bulb will last longer."

"That's right," says the engineer. "Light bulbs behave as you expect; they wear out over time, so the longer they've been in use, the sooner they burn out, on average."

"However," says the obstetrician, "not everything works that way. For example, most often, pregnancy lasts 39 or 40 weeks. Today I saw three patients who are all pregnant; the first is at the beginning of week 39, the second is at the beginning of week 40, and the third is at the beginning of week 41. Which one do you think will deliver her baby first?"

Now you are sure it's a trick question, but just to play along, you say, "The third patient is likely to deliver first."

The obstetrician says, "No, the remaining duration of the three pregnancies is nearly the same, about four days. Even taking medical intervention into account, all three have the same chance of delivering first."

"That's surprising," says the oncologist. "But in my field things are even stranger. For example, today I saw two patients with glioblastoma, which is a kind of brain cancer. They are about the same age, and the stage of their cancers is about the same, but one of them was diagnosed a week ago and one was diagnosed a year ago. Unfortunately, the average survival time after diagnosis is only about a year. So you probably expect the first patient to live longer."

By now you know better than to guess, so you wait for the answer. The oncologist explains that many patients with glioblastoma live only a few months after diagnosis. So, it turns out, a patient who survives one year after diagnosis is then *more* likely to survive a second year.

Based on this conversation, we can see that there are three ways survival times can go.

- Many things wear out over time, like light bulbs, so we expect something new to last longer than something old.
- But there are some situations, like patients after a cancer diagnosis, that are the other way around: the longer someone has survived, the longer we expect them to survive.
- And there are some situations, like women expecting babies, where the average remaining time doesn't change, at least for a while.

In this chapter I'll demonstrate and explain each of these effects, starting with light bulbs.

LIGHT BULBS

As your engineer friend asked at lunch, would you rather have a new light bulb or one that has been used for 1000 hours? To answer this question, I'll use data from an experiment run in 2007 by research-

ers at Banaras Hindu University in India. They installed 50 new incandescent light bulbs in a large rectangular area, turned them on,
and left them burning continuously until the last bulb expired 2568
hours later. During this period, which is more than three months,
they checked the bulbs every 12 hours and recorded the number
that had expired.

The following figure shows the distribution of survival times for
these light bulbs, plotted as a cumulative distribution function (CDF),
along with a Gaussian model. The shaded area shows how much variation we expect from the Gaussian model. Except for one unusually
long-lasting bulb, the lifetimes fall within the bounds, which shows
that the data are consistent with the model.

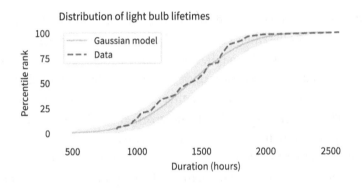

The researchers who collected this data explain why we might expect the distribution to be Gaussian. While a light bulb with a tungsten filament burns, "the evaporation of tungsten atoms proceeds
and the hot filament gets thinner," until it breaks and the lamp fails.
Using a statistical model of this evaporation, they show that the resulting distribution of lifetimes is Gaussian.

In this dataset, the average lifetime for a new light bulb is about
1414 hours. For a bulb that has been used for 1000 hours, the average
lifetime is higher, about 1495 hours; however, since it has already
burned 1000 hours, its average remaining lifetime is only 495 hours.
So we would rather have the new bulb.

We can do the same calculation for a range of elapsed times from
0 to 2568 hours (the lifespan of the oldest bulb). At each point in

time, t, we can compute the average lifetime for bulbs that survive past t and the average remaining lifetime we expect. The following figure shows the result.

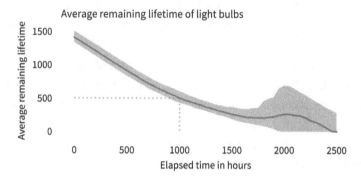

The x-axis shows elapsed times since the installation of a hypothetical light bulb. The y-axis shows the average remaining lifetime. In this example, a bulb that has been burning for 1000 hours is expected to last about 495 hours more, as indicated by the dotted lines.

Between zero and 1700 hours, this curve is consistent with intuition: The longer the bulb has been burning, the sooner we expect it to expire. Then, between 1700 and 2000 hours, it gets better! If we understand that the filament of a light bulb gets thinner as it burns, does this curve mean that sometimes the evaporation process goes in reverse and the filament gets thicker? That seems unlikely.

The reason for this reversal is the presence of one bulb that lasted 2568 hours, almost twice the average. So one way to think about what's going on is that there are two kinds of bulbs: ordinary bulbs and Super Bulbs. The longer a bulb lasts, the more likely it is to be a Super Bulb. And the more likely it is to be a Super Bulb, the longer we expect it to last. Nevertheless, a new bulb generally lasts longer than a used bulb.

ANY DAY NOW

Now let's consider the question posed by the imaginary obstetrician at lunch. Suppose you visit a maternity ward and meet women who are starting their 39th, 40th, and 41st weeks of pregnancy. Which one do you think will deliver first?

To answer this question, we need to know the distribution of gestation times, which we can get from the National Survey of Family Growth (NSFG), which was the source of the birth weights in chapter 4. I gathered data collected between 2002 and 2017, which includes information about 43,939 live births. The following figure shows the distribution of their durations (except for the 1% of babies born prior to 28 weeks). About 41% of these births were during the 39th week of pregnancy, and another 18% during the 40th week.

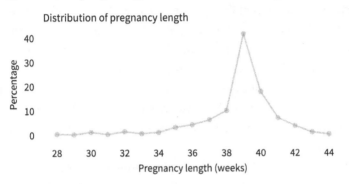

By now you probably expect me to say that this distribution follows a Gaussian or lognormal model, but it doesn't. It is not symmetric like a Gaussian distribution or skewed to the right like most lognormal distributions. And the most common value is more common than we would expect from either model. As we've seen several times now, nature is under no obligation to follow simple rules; this distribution is what it is. Nevertheless, we can use it to compute the average remaining time as a function of elapsed time, as shown in the following figure.

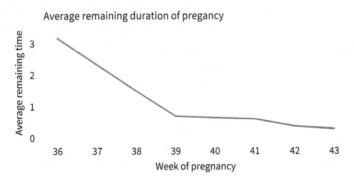

Between weeks 36 and 39, the curve behaves as we expect: as time goes on, we get closer to the finish line. For example, at the beginning of week 36, the average remaining time is 3.2 weeks. At the beginning of week 37, it is down to 2.3 weeks. So far, so good: a week goes by and the finish line is just about one week closer.

But then, cruelly, the curve levels off. At the beginning of week 39, the average remaining time is 0.68 weeks, so the end seems near. But if we get to the beginning of week 40 and the baby has not been born, the average remaining time is 0.63 weeks. A week has passed and the finish line is only eight hours closer. And if we get to the beginning of week 41 and the baby has not been born, the average remaining time is 0.59 weeks. Another week has passed and the finish line is only seven hours closer! These differences are so small that they are only measurable because we have a large sample; for practical purposes, the expected remaining time is essentially unchanged for more than two weeks.

This is why if you ask a doctor how long it will be until a baby is born, they generally say something noncommittal like "Any day now." As infuriating as this answer might be, it sums up the situation pretty well, and it is probably kinder than the truth.

CANCER SURVIVAL TIMES

Finally, let's consider the surprising result reported by the imaginary oncologist at lunch: for many cancers, a patient who has survived a year after diagnosis is expected to live longer than someone who has just been diagnosed.

To demonstrate and explain this result, I'll use data from the the Surveillance, Epidemiology, and End Results (SEER) program, which is run by the US National Institutes of Health (NIH). Starting in 1973, SEER has collected data on cancer cases from registries in several locations in the United States. In the most recent datasets, these registries cover about one-third of the US population.

From the SEER data, I selected the 16,202 cases of glioblastoma, diagnosed between 2000 and 2016, where the survival times are available. We can use this data to infer the distribution of survival times, but first we have to deal with a statistical obstacle: some of the patients are still alive, or were alive the last time they appeared in a registry.

For these cases, the time until their death is unknown, which is a good thing. However, in order to work with data like this, we need a special method, called Kaplan-Meier estimation, to compute the distribution of lifetimes. The following figure shows the result on a log scale, plotted in the familiar form of a CDF and in a new form called a survival curve.

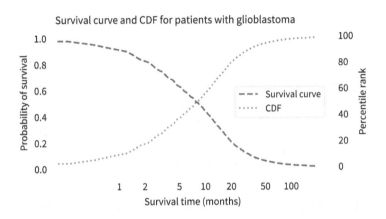

The survival curve shows the probability of survival past a given time on a scale from 0 to 1. It is the complement of the CDF, so as the CDF increases from left to right, the survival curve decreases. The two curves contain the same information; the only reason to use one or the other is convention. Survival curves are used more often in medicine and reliability engineering; CDFs, in many other fields. One thing that is apparent—from either curve—is that glioblastoma is a serious diagnosis. The median survival time after diagnosis is less than nine months, and only about 16% of patients survive more than two years.

Please keep in mind that this curve lumps together people of different ages with different health conditions, diagnosed at different stages of disease over a period of about 16 years. Survival times depend on all of these factors, so this curve does not provide a prognosis for any specific patient. In particular, as treatment has gradually improved, the prognosis is better for someone with a more recent diagnosis. If you or someone you know is diagnosed with glioblastoma, you should get a prognosis from a doctor, based on specifics of the case, not from aggregated data in a book demonstrating basic statistical methods.

As we did with light bulbs and pregnancy lengths, we can use this distribution to compute the average remaining survival time for patients at each time after diagnosis. The following figure shows the result.

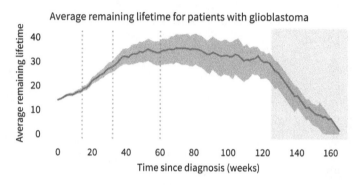

At the time of diagnosis, the average survival time is about 14 months. That is certainly a bleak prognosis, but there is some good news to follow. If a patient survives the first 14 months, we expect them to survive another 18 months, on average. If they survive those 18 months, for a total of 32, we expect them to survive another 28 months. And if they survive those 28 months, for a total of 60 months (five years), we expect them to survive another 35 months (almost three years). The vertical lines indicate these milestones. It's like running a race where the finish line keeps moving, and the farther you go, the faster it retreats.

After 60 months, the curve levels off, which means that the expected remaining survival time is constant. Finally, after 120 months (10 years), it starts to decline. However, we should not take this part of the curve too seriously, which is why I grayed it out. Statistically, it is based on a small number of cases. Also, most people diagnosed with glioblastoma are over 60 (the median in this dataset is 64). Ten years after diagnosis, they are even older, so some of the decline in this part of the curve is the result of deaths from other causes.

This example shows how cancer patients are different from light bulbs. In general, we expect a new light bulb to last longer than an old one; this property is called "new better than used in expectation," abbreviated NBUE. The term "in expectation" is another way of saying "on average." But for some cancers, we expect a patient who has survived some time after diagnosis to live longer. This property is called "new worse than used in expectation," abbreviated NWUE.

The idea that something new is worse than something used is contrary to our experience of things in the world that wear out over time. It is also contrary to the behavior of a Gaussian distribution. For example, if you hear that the average survival time after diagnosis is 14 months, you might imagine a Gaussian distribution where 14 months is the most common value and an equal number of patients live more or less than 14 months. But that would be a very misleading picture.

To show how bad it would be, I chose a Gaussian distribution that matches the distribution of survival times as well as possible—which is not very well—and used it to compute average remaining survival times. The following figure shows the result, along with the actual averages.

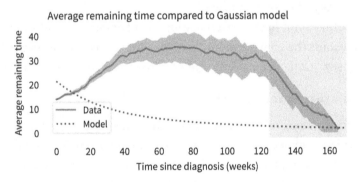

With the Gaussian model, the average remaining survival time starts around 20 months, drops quickly at first, and levels off around 5 months. So it behaves nothing like the actual averages. If your mental model of the distribution is Gaussian, you would seriously misunderstand the situation!

On the other hand, if your mental model of the distribution is lognormal, you would get it about right. To demonstrate, I chose a lognormal distribution that fits the actual distribution of survival times and used it to compute average remaining lifetimes. The following figure shows the result.

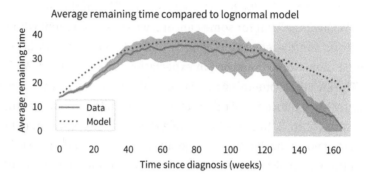

During the first 24 months, the model is a little too optimistic, and after 120 months it is much too optimistic. But the lognormal model gets the shape of the curve right: if your mental model of the distribution is lognormal, you would have a reasonably accurate understanding of the situation. And you would understand why a patient who has survived three years is likely to live longer than a patient who has just been diagnosed.

At this point in the imaginary lunch, a demographer at a nearby table joins the conversation. "Actually, it's not just cancer patients," they say. "Until recently, every person born was better used than new."

LIFE EXPECTANCY AT BIRTH

In 2012, a team of demographers at the University of Southern California estimated life expectancy for people born in Sweden in the early 1800s and 1900s. They chose Sweden because it "has the deepest historical record of high-quality [demographic] data." For ages from 0 to 91 years, they estimated the mortality rate, which is the fraction of people at each age who die. The following figure shows the results for two cohorts: people born between 1800 and 1810 and people born between 1905 and 1915.

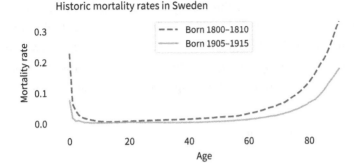

Historic mortality rates in Sweden

The most notable feature of these curves is their shape; they are called "bathtub curves" because they drop off steeply on one side and increase gradually on the other, like the cross section of a bathtub. The other notable feature is that mortality rates were lower for the later cohort at every age. For example:

- On the left side, mortality rates are highest during the first year. In the earlier cohort (1800–1810), about 23% died before their first birthday; in the later cohort (1905–1915), it was about 7%.
- Both curves drop quickly to a minimum in young adulthood. In the earlier cohort, the mortality rate at age 14 is about 5 per 1000. In the later cohort, it is about 2 per 1000.
- On the right side, mortality rates increase with age. For example, of people in the earlier cohort who made it to age 80, about 13% died at age 80. In the later cohort, the same mortality rate was about 6%.

We can use these curves to compute "life expectancy at birth." For people born around 1800, the average lifespan was about 36 years; for people born around 1900, it was 66 years. So that's a substantial change. However, life expectancy at birth can be misleading. When people hear that the average lifespan was 36 years, they might imagine a Gaussian distribution where many people die near age 36 and few people live much longer than that. But that's not accurate.

For someone born in 1800, if they made it to age 36, they would expect to live another 29 years, for a total of 65. And if they made it to 65, they would expect to live another 11 years, for a total of 76, which is more than twice the average. The reason for this counterintuitive behavior is child mortality. If child mortality is high, life expectancy at birth is low. But if someone survives the first few years, their life expectancy goes up. The following figure shows life expectancy as a function of age for people born in Sweden around 1800 and 1905.

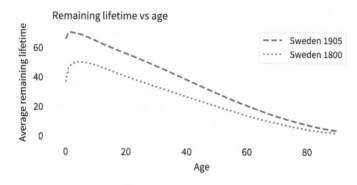

In both cohorts, used was better than new, at least for the first few years of life. For someone born around 1800, life expectancy at birth was only 36. But if they survived the first five years, they would expect to live another 50 years, for a total of 55. Similarly, for someone born in 1905, life expectancy at birth was 66 years. But if they survived the first two years, they would expect to live another 70, for a total of 72.

CHILD MORTALITY

Fortunately, child mortality has decreased since 1900. The following figure shows the percentage of children who die before age five for four geographical regions, from 1900 to 2019. These data were combined from several sources by Gapminder, a foundation based in Sweden that "promotes sustainable global development [. . .] by increased use and understanding of statistics."

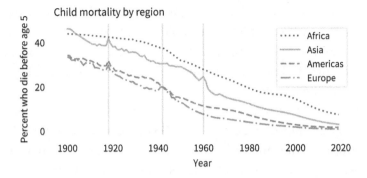

In every region, child mortality has decreased consistently and substantially. The only exceptions are indicated by the vertical lines: the 1918 influenza pandemic, which visibly affected Asia, the Americas, and Europe; World War II in Europe (1939–1945); and the Great Leap Forward in China (1958–1962). In every case, these exceptions did not affect the long-term trend.

Although there is more work to do, especially in Africa, child mortality is substantially lower now, in every region of the world, than in 1900. As a result most people now are better new than used. To demonstrate this change, I collected recent mortality data from the Global Health Observatory of the World Health Organization (WHO). For people born in 2019, we don't know what their future lifetimes will be, but we can estimate it if we assume that the mortality rate in each age group will not change over their lifetimes. Based on that simplification, the following figure shows average remaining lifetime as a function of age for Sweden and Nigeria in 2019, compared to Sweden in 1905.

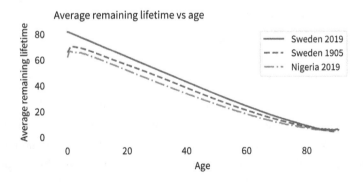

Since 1905, Sweden has continued to make progress; life expectancy at every age is higher in 2019 than in 1905. And Swedes now have the NBUE property. Their life expectancy at birth is about 82 years, and it declines consistently over their lives, like the life expectancy of a light bulb.

Unfortunately, Nigeria has one of the highest rates of child mortality in the world: in 2019, almost 8% of babies died in their first year of life. After that, they are briefly better used than new: life expectancy at birth is about 62 years; however, a baby who survives the first year will live another 65 years, on average. Going forward, I hope we continue to reduce child mortality in every region; if we do, soon every person born will be better new than used. Or maybe we can do even better than that.

THE IMMORTAL SWEDE

In the previous section, I showed that child mortality declined quickly in the past century, and we saw the effect of these changes on life expectancy. In this section we'll look more closely at adult mortality. Going back to the data from Sweden, the following figure shows the mortality rate for each age group, updated every 10 years from 2000 to 2019.

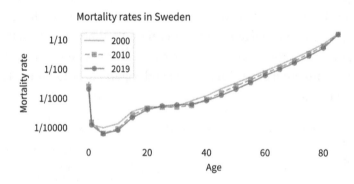

Notice that the *y*-axis is on a log scale, so the differences between age groups are orders of magnitude. In 2019, the mortality rate for people over 85 was about one in 10. For people between 65 and 69 it was one in 100; between 40 and 45, it was one in 1000; and for people between 10 and 14, it was less than one in 10,000. This figure shows that there are four phases of mortality:

- In the first year of life, mortality is still quite high. Infants have about the same mortality rate as 50-year-olds.
- Among children between 1 and 19, mortality is at its lowest.
- Among adults from 20 to 35, it is low and almost constant.
- After age 35, it increases at a constant rate.

The straight-line increase after age 35 was described by Benjamin Gompertz in 1825, so this phenomenon is called the Gompertz Law. It is an empirical law, which is to say that it names a pattern we see in nature, but at this point we don't have an explanation of why it's true, or whether it is certain to be true in the future. Nevertheless, the data in this example fall in a remarkably straight line.

The previous figure also shows that mortality rates decreased between 2000 and 2019 in almost every age group. If we zoom in on the age range from 40 to 80, as shown in the following figure, we can see the changes in adult mortality more clearly.

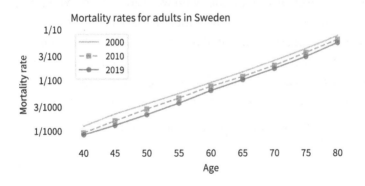

In these age groups, the decreases in mortality have been remarkably consistent. By fitting a model to this data, we can estimate the rate of change as a function of both age and time. According to the model, as you move from one age group to the next, your mortality rate increases by about 11% per year. At the same time, the mortality rate in every age group decreases by about 2% per year.

These results imply that the life expectancies we computed in the previous section are too pessimistic, because they take into account only the first effect—the increase with age—and not the second—the decrease over time. So let's see what happens if we include the second effect as well—that is, if we assume that mortality rates will continue to decrease. The following figure shows the actual mortality rates for 2000 and 2019 again, along with predictions for 2040 and 2060.

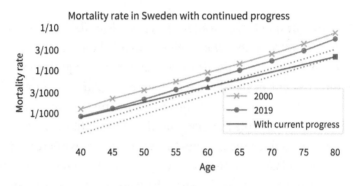

The line labeled "With current progress" indicates the mortality rates we expect for Swedes who were 40 in 2020. When they are 60, it will be 2040, so we expect them to have the mortality rate of a 60-year-old in 2040, indicated by a triangle. And when they are 80, it will be 2060, so we expect them to have the mortality rate of an 80-year-old in 2060, indicated by a square. We can use these mortality rates to compute survival curves, as shown in the following figure.

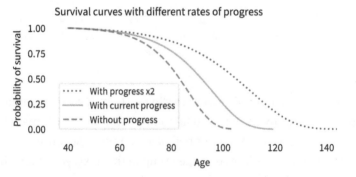

The dashed line on the left shows the survival curve we expect if there is no further decrease in mortality rates; in that scenario, life expectancy at age 40 is 82 years, and the probability of living to 100 is only 1.4%. The solid line shows the survival curve if mortality continues to decrease at the current rate; in that case, life expectancy for the same 40-year-old is 90 years, and the chance of living to 100 is 25%. Finally, the dotted line on the right shows the survival curve if mortality decreases at twice the current rate. Life expectancy at age 40 would be 102, and the probability of living to 100 would be 60%.

Of course, we don't expect the pace of progress to double overnight, but we could get there eventually. In a recent survey, demographers in Denmark and the United States listed possible sources of accelerating progress, including

- prevention and treatment of infectious disease and prevention of future pandemics;
- reduction in lifestyle risk factors, like obesity and drug abuse (and I would add improvement in suicide prevention);
- prevention and treatment of cancer, possibly including immune therapies and nanotechnology;
- precision medicine that uses individual genetic information to choose effective treatment, and CRISPR treatment for genetic conditions;
- technology for reconstructing and regenerating tissues and organs;
- and possibly ways to slow the rate of biological aging.

To me it seems plausible that the rate of progress could double in the near future. But even then, everyone would die eventually. Maybe we can do better. The following figure shows the survival curve under the assumption that mortality rates, starting in 2019, decrease at four times the current rate.

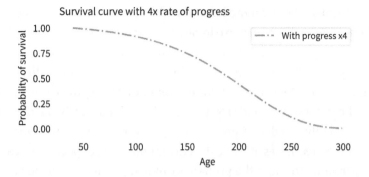

In this scenario, some people live to be 300 years old! However, even with these optimistic assumptions, the shape of the curve is similar to what we get with slower rates of progress. And, as shown in the following figure, average remaining lifetimes decrease with almost the same slope.

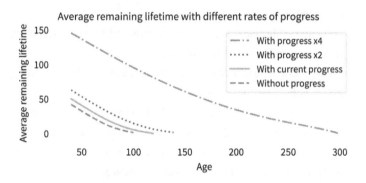

With faster progress, people live longer, but they still have the NBUE property: as each year passes, they get closer to the grave, on average. However, something remarkable happens if progress accelerates by a factor of 4.9. At that speed, the increase in mortality due to aging by one year is exactly offset by the decrease due to progress. That means that the probability of dying is the same from one year to the next, forever. The result is a survival curve that looks like this:

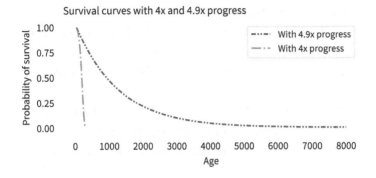

Survival curves with 4x and 4.9x progress

The difference between 4 and 4.9 is qualitative. At a factor of 4, the survival curve plummets toward an inevitable end; at a factor of 4.9, it extends out to ages that are beyond biblical.

With progress at this rate, half the population lives to be 879, about 20% live to be 2000, and 4% live to be 4000. Out of seven billion people, we expect the oldest to be 29,000 years old. But the strangest part is the average *remaining* lifetime. In the 4.9 scenario, life expectancy at birth is 1268 years. But if you survive 1268 years, your average remaining lifetime is still 1268 years. And if you survive those years, your expected remaining lifetime is *still* 1268 years, and so on, forever.

Every year, your risk of dying goes up because you are older, but it goes down by just as much due to progress. People in this scenario are not immortal; each year about eight out of 10,000 die. But as each year passes, you take one step toward the grave, and the grave takes one step away from you. In the near future, if child mortality continues to decrease, everyone will be better new than used. But eventually, if adult mortality decreases fast enough, everyone will have the same remaining time, on average: new, used, or in between.

SOURCES AND RELATED READING

- The paper about light bulbs is "Renewal Rate of Filament Lamps" [79].
- The data on pregnancy durations is from the National Survey of Family Growth, available from the website of the CDC [84].
- The data on cancer survival times is from the Surveillance, Epidemiology, and End Results (SEER) Program of the National Cancer Institute [119].

- The data on childhood mortality is from Gapminder [18].
- Historical mortality data in Sweden is from "Early Cohort Mortality Predicts the Rate of Aging in the Cohort" [12].
- More recent mortality data is from the WHO Global Health Observatory [69].
- The paper on potential future reductions in mortality rates is "Demographic Perspectives on the Rise of Longevity" [129].
- A recent book on future developments in longevity is Steven Johnson's *Extra Life* [55].

CHAPTER 6

JUMPING TO CONCLUSIONS

In 1986, Michael Anthony Jerome "Spud" Webb was the shortest player in the National Basketball Association (NBA); he was five feet six inches (168 cm). Nevertheless, that year he won the NBA Slam Dunk Contest after successfully executing the elevator two-handed double pump dunk and the reverse two-handed strawberry jam. This performance was possible because Webb's vertical leap was at least 42 inches (107 cm), bringing the top of his head to nine feet (275 cm) and his outstretched hand well above the basketball rim at 10 feet (305 cm).

During the same year, I played intercollegiate volleyball at the Massachusetts Institute of Technology (MIT). Although the other players and I were less elite athletes than Spud Webb, I noticed a similar relationship between height and vertical leap. Consistently, the shorter players on the team jumped higher than the taller players. At the time, I wondered if there might be a biokinetic reason for the relationship: maybe shorter legs provide some kind of mechanical advantage. It turns out that there is no such reason; according to a 2003 paper on physical characteristics that predict jumping ability, there is "no significant relationship" between body height and vertical leap.

By now, you have probably figured out the error in my thinking. In order to play competitive volleyball, it helps to be tall. If you are not tall, it helps if you can jump. If you are not tall and can't jump, you probably don't play competitive volleyball, you don't play in the NBA, and you definitely don't win a slam dunk contest.

In the general population, there is no correlation between height and jumping ability, but intercollegiate athletes are not a representative sample of the population, and elite athletes even less so. They have been selected based on their height and jumping ability, and the selection process creates the relationship between these traits. This phenomenon is called Berkson's paradox after a researcher who wrote about it in 1946. I'll present his findings later, but first let's look at another example from college.

MATH AND VERBAL SKILLS

Among students at a given college or university, do you think math and verbal skills are correlated, anti-correlated, or uncorrelated? In other words, if someone is above average at one of these skills, would you expect them to be above average on the other, or below average, or do you think they are unrelated? To answer this question, I will use data from the National Longitudinal Survey of Youth 1997 (NLSY97), which "follows the lives of a sample of 8984 American youth born between 1980–84." The public dataset includes the participants' scores on several standardized tests, including the tests most often used in college admissions, the SAT and ACT.

In this cohort, about 1400 participants took the SAT. Their average scores were 502 on the verbal section and 503 on the math section, both close to the national average, which is calibrated to be 500. The standard deviation of their scores was 108 on the verbal section and 110 on the math section, both a little higher than the overall standard deviation, which is calibrated to be 100.

To show how the scores are related, here's a scatter plot showing a data point for each participant, with verbal scores on the horizontal axis and math scores on the vertical:

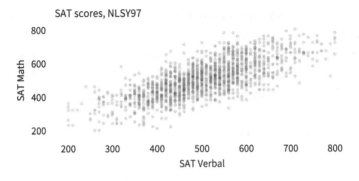

People who do well on one section tend to do well on the other. The correlation between scores is 0.73, so someone who is one standard deviation above the mean on the verbal test is expected to be 0.73 standard deviations above the mean on the math test. For example, if you select people whose verbal score is near 600, which is 100 points above the mean, the average of their math scores is 570, about 70 points above the mean.

This is what we would see in a random sample of people who took the SAT: a strong correlation between verbal and mathematical ability. But colleges don't select their students at random, so the ones we find on any particular campus are not a representative sample. To see what effect the selection process has, let's think about a few kinds of colleges.

ELITE UNIVERSITY

First, consider an imaginary institution of higher learning called Elite University. To be admitted to EU, let's suppose, your total SAT score (sum of the verbal and math scores) has to be 1320 or higher. If we select participants from the NLSY who meet this requirement, their average score on both sections is about 700, the standard deviation is about 50, and the correlation of the two scores is −0.33. So if someone is one standard deviation above the EU mean on one test, we expect them to be one-third of a standard deviation *below* the mean on the other, on average. For example, if you meet an EU student who got a 760 on the verbal section (60 points above the EU mean), you would expect them to get a 680 on the math section (20 points below the mean).

In the population of test takers, the correlation between scores is positive. But in the population of EU students, the correlation is negative. The following figure shows how this happens.

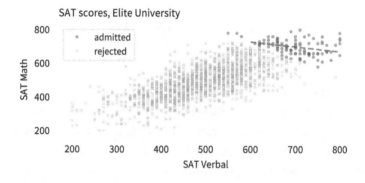

The circles in the upper right show students who meet the requirement for admission to Elite University. The dashed line shows the average math score for students at each level of verbal score. If you meet someone at EU with a relatively low verbal score, you expect them to have a high math score. Why? Because otherwise, they would not be at Elite University. And conversely, if you meet someone with a relatively low math score, they must have a high verbal score.

I suspect that this effect colors the perception of students at actual elite universities. Among the students they meet, they are likely to encounter some who are unusually good at math, some whose verbal skills are exceptional, and only a few who excel at both.

These interactions might partly explain a phenomenon described by C. P. Snow in his famous lecture, "The Two Cultures":

A good many times I have been present at gatherings of people who, by the standards of the traditional culture, are thought highly educated and who have with considerable gusto been expressing their incredulity at the illiteracy of scientists. Once or twice I have been provoked and have asked the company how many of them could describe the Second Law of Thermodynamics. The response was cold: it was also negative. Yet I was asking something which is about the scientific equivalent of "Have you read a work of Shakespeare's?"

I now believe that if I had asked an even simpler question—such as, What do you mean by mass, or acceleration, which is the scientific equivalent of saying, "Can you read?"—not more than one in ten of the highly educated would have felt that I was speaking the same language.

In the general population, math and verbal abilities are highly correlated, but as people become "highly educated," they are also highly selected. And as we'll see in the next section, the more stringently they are selected, the more these abilities will seem to be anti-correlated.

LESS ELITE, MORE CORRELATED

Of course, not all colleges require a total SAT score of 1320. Most are less selective, and some don't consider standardized test scores at all. So let's see what happens to the correlation between math and verbal scores as we vary admission requirements. As in the previous example, I'll use a model of the selection process that will annoy anyone who works in college admission:

- Suppose that every college has a simple threshold for the total SAT score. Any applicant who exceeds that threshold will be offered admission.
- And suppose that the students who accept the offer are a random sample from the ones who are admitted.

The following figure shows a range of thresholds on the horizontal axis, from a total score of 700 up to 1400. On the vertical axis, it shows the correlation between math and verbal scores among the students whose scores exceed the threshold, based on the NLSY data.

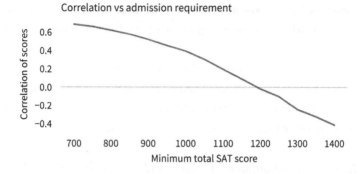

At a college that requires a total score of 700, math and verbal scores are positively correlated. At a selective college where the threshold is around 1200, the correlation is close to zero. And at an elite school where the threshold is over 1300, the correlation is negative.

The examples so far are based on an unrealistic model of the college admission process. We can make it slightly more realistic by adding one more factor.

SECONDTIER COLLEGE

Suppose at Secondtier College (it's pronounced "se-con-tee-ay"), a student with a total score of 1200 or more is admitted, but a student with 1300 or more is likely to go somewhere else. Using data from the NLSY again, the following figure shows the verbal and math scores for applicants in three groups: rejected, enrolled, and the ones who were accepted but went somewhere else. Among the students who enrolled, the correlation between math and verbal scores is strongly negative, about –0.84.

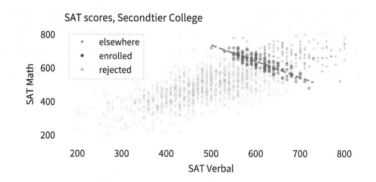

At Secondtier, if you meet a student who got 650 on the verbal section (about 50 points above the mean), you should expect them to have gotten 590 on the verbal section (about 40 points below the mean). So we have the answer to the question I posed: depending on how colleges select students, math and verbal scores might be correlated, unrelated, or anti-correlated.

BERKSON'S PARADOX IN HOSPITAL DATA

The examples we've seen so far are interesting, I hope, but they might not be important. If you are wrong about the relationship between height and jumping ability or between math and verbal skills, the consequences of your mistake are limited. But Berkson's paradox is not limited to sports and standardized tests. It appears in many domains where the consequences of being wrong are more serious, including the field where it was first widely recognized: medicine.

Berkson's paradox is named for Joseph Berkson, who led the Division of Biometry and Medical Statistics at the Mayo Clinic in Rochester, Minnesota. In 1946, he wrote a paper pointing out the danger of using patients in a clinic or hospital as a sample. As an example, he uses the relationship between cholecystic disease (inflammation of the gall bladder) and diabetes. At the time, these conditions were thought to be related so that, "in certain medical circles, the gall bladder was being removed as a treatment for diabetes." His tone suggests what he thinks of these "medical circles."

Berkson shows that the apparent relationship between the two conditions might be the result of using hospital patients as a sample. To demonstrate the point, he generates a simulated population where 1% have diabetes, 3% have cholecystitis, and the conditions are unrelated; that is, having one condition does not, in fact, change the probability of having the other. Then he simulates what happens if we use hospital patients as a sample. He assumes that people with either condition are more likely to go to the hospital, and therefore more likely to appear in the hospital sample, compared to people with neither condition. In the hospital sample, he finds that there is a negative correlation between the conditions; that is, people with cholecystitis are less likely to have diabetes.

I'll demonstrate with a simplified version of Berkson's experiment. Consider a population of one million people where 1% have diabetes, 3% have cholecystitis, and the conditions are unrelated. We'll assume that someone with either condition has a 5% chance of appearing in the hospital sample, and someone with neither condition has a 1% chance.

If we simulate this sampling process, we can compute the number of people in the hospital sample with and without cholecystitis, and with and without diabetes. The following table shows the results in a form Berkson calls a "fourfold table."

	C	No C	Total
D	29	485	514
No D	1485	9603	11,088

The columns labeled "C" and "no C" represent hospital patients with and without cholecystitis; the rows labeled "D" and "no D" represent patients with and without diabetes. The results indicate that, from a population of one million people:

- we expect a total of 514 people with diabetes (top row) to appear in a sample of hospital patients; of them, we expect 29 to also have cholecystitis; and
- we expect 11,088 people without diabetes (bottom row) to appear in the sample; of them, we expect 1485 to have cholecystitis.

If we compute the rate in each row, we find that

- in the group with diabetes, about 5.6% have cholecystitis, and
- in the group without diabetes, about 13% have cholecystitis.

So in the sample, the correlation between the conditions is negative: if you have diabetes, you are substantially less likely to have cholecystitis. But we know that there is no such relationship in the simulated population, because we designed it that way. So why does this relationship appear in the hospital sample?

Here are the pieces of the explanation:

- People with either condition are more likely to go to the hospital, compared to people with neither condition.
- If you find someone at the hospital who has cholecystitis, that's likely to be the reason they are in the hospital.
- If they *don't* have cholecystitis, they are probably there for another reason, so they are more likely to have diabetes.

Berkson showed that it can also work the other way; under different assumptions, a hospital sample can show a *positive* correlation between conditions, even if we know there is no such relationship in the general population.

He concludes that "there does not appear to be any ready way of correcting the spurious correlation existing in the hospital population," which is "the result merely of the ordinary compounding of independent probabilities." In other words, even if you know that an observed correlation might be caused by Berkson's paradox, the only way to correct it, in general, is to collect a better sample.

BERKSON AND COVID-19

This problem has not gone away. A 2020 paper argues, right in the title, that Berkson's paradox "undermines our understanding of COVID-19 disease risk and severity." The authors, who include researchers from the University of Bristol in the United Kingdom and the Norwegian University of Science and Technology, point out the problem with using samples from hospital patients, as Berkson did, and also "people tested for active infection, or people who volunteered." Patterns we observe in these groups are not always consistent with patterns in the general population.

As an example, they consider the hypothesis that health care workers might be at higher risk for severe COVID disease. That could be the case if COVID exposures in hospitals and clinics involve higher viral loads, compared to exposures in other environments. To test this hypothesis, we might start with a sample of people who have tested positive for COVID and compare the frequency of severe disease among health care workers and others.

But there's a problem with this approach: not everyone is equally likely to get tested. In many places, health care workers are tested regularly whether they have symptoms or not, while people in lower-risk groups are tested only if they have symptoms. As a result, the sample of people who test positive includes health care workers in the entire range from severe disease to mild symptoms to no symptoms at all; but for other people, it includes only cases that are severe enough to warrant testing. In the sample, COVID disease might seem milder for health care workers than for others, even if, in the general population, it is more severe for health care workers on average.

This scenario is hypothetical, but the authors go on to explore data from the UK BioBank, which is "a large-scale biomedical database and research resource, containing in-depth genetic and health information from half a million UK participants." To see how pervasive Berkson's paradox might be, the authors selected 486,967 BioBank participants from April 2020. For each participant, the dataset reports 2556 traits, including demographic information like age, sex, and socioeconomic status; medical history and risk factors; physical and psychological measurements; and behavior and nutrition.

Of these traits, they found 811 that were statistically associated with the probability of being tested for COVID, and many of these associations were strong enough to induce Berkson's paradox. The traits they identified included occupation (as in the example of the health care workers), ethnicity and race, general health, and place of residence, as well as less obvious factors like internet access and scientific interest. As a result, it might be impossible to estimate accurately the effect of any of these traits on disease severity using only a sample of people who tested positive for COVID.

The authors conclude, "It is difficult to know the extent of sample selection, and even if that were known, it cannot be proven that it has been fully accounted for by any method. Results from samples that are likely not representative of the target population should be treated with caution by scientists and policy makers."

BERKSON AND PSYCHOLOGY

Berkson's paradox also affects the diagnosis of some psychological conditions. For example, the Hamilton Rating Scale for Depression (HRSD) is a survey used to diagnose depression and characterize its severity. It includes 17 questions, some on a three-point scale and some on a five-point scale. The interpretation of the results depends on the sum of the points from each question. On one version of the questionnaire, a total between 14 and 18 indicates a diagnosis of mild depression; a total above 18 indicates severe depression.

Although rating scales like this are useful for diagnosis and evaluation of treatment, they are vulnerable to Berkson's paradox. In a 2019 paper, researchers from the University of Amsterdam and Leiden University in the Netherlands explain why. As an example, they consider the first two questions on the HRSD, which ask about feelings of sadness and feelings of guilt. In the general population, responses to these questions might be positively correlated, negatively correlated, or not correlated at all.

But if we select people whose total score indicates depression, the scores on these items are likely to be negatively correlated, especially at the intermediate levels of severity. For example, someone with mild depression who reports high levels of sadness is likely to have low levels of guilt, because if they had low levels of both, they would probably not have depression, and if they had high levels of both, it would probably not be mild.

This example considers two of the 17 questions, but the same problem affects the correlation between any two items. The authors conclude that "Berkson's bias is a considerable and under-appreciated problem" in clinical psychology.

BERKSON AND YOU

Now that you know about Berkson's paradox, you might start to notice it in your everyday life. Does it seem like restaurants in bad locations have really good food? Well, maybe they do, compared to restaurants in good locations, because they have to. When a restaurant opens, the quality of the food might be unrelated to the location. But if a bad restaurant opens in a bad location, it probably doesn't last very long. If it survives long enough for you to try it, it's likely to be good.

When you get there, you should order the least appetizing thing on the menu, according to Tyler Cowen, a professor of economics at George Mason University. Why? If it sounds good, people will order it, so it doesn't have to be good. If it sounds bad, and it is bad, it wouldn't be on the menu. So if it sounds bad, it has to be good.

Or here's an example from Jordan Ellenberg's book *How Not to Be Wrong*. Among people you have dated, does it seem like the attractive ones were meaner, and the nice ones were less attractive? That might not be true in the general population, but the people you've dated are not a random sample of the general population, we assume. If someone is unattractive and mean, you will not choose to date them. And if someone is attractive and nice, they are less likely to be available. So, among what's left, you are more likely to get one or the other, and it's not easy to find both.

Does it seem like when a movie is based on a book, the movie is not as good? In reality, there might be a positive correlation between the quality of a movie and the quality of the book it is based on. But if someone makes a bad movie from a bad book, you've probably never heard of it. And if someone makes a bad movie from a book you love, it's the first example that comes to mind. So in the population of movies you are aware of, it might seem like there's a negative correlation between the qualities of the book and the movie.

In summary, for a perfect date, find someone unattractive, take them to a restaurant in a strip mall, order something that sounds awful, and go to a movie based on a bad book.

SOURCES AND RELATED READING

- Spud Webb's performance in the 1986 NBA Slam Dunk Contest is documented by Wikipedia [116] and viewable on internet video sites.

- The paper on factors that predict jumping ability is "Physical Characteristics That Predict Vertical Jump Performance in Recreational Male Athletes" [27].
- Data from the National Longitudinal Survey of Youth is available from the US Bureau of Labor Statistics [83].
- C. P. Snow's lecture, "The Two Cultures," was published as a book in 1959 [114].
- The paper on Berkson's paradox and COVID-19 is "Collider Bias Undermines Our Understanding of COVID-19 Disease Risk and Severity" [50].
- The paper on the consequences of Berkson's paradox in psychological testing is "Psychological Networks in Clinical Populations" [28].
- Lionel Page collected examples of Berkson's paradox in a Twitter thread [90].
- Jordan Ellenberg wrote about Berkson's paradox in *How Not to Be Wrong* [38].
- Richard McElreath also suggests that restaurants in bad locations tend to have good food [77].
- Tyler Cowen suggests you should order the least appetizing thing on the menu in "Six Rules for Dining Out" [25]. In fact, the article is a catalog of Berkson paradoxes.
- In a Numberphile video, Hannah Fry suggests that the apparent relationship between the quality of movies and the books is an example of Berkson's bias [87].

CHAPTER 7

CAUSATION, COLLISION, AND CONFUSION

The low-birthweight paradox was born in 1971, when Jacob Yerushalmy, a researcher at the University of California, Berkeley, published "The Relationship of Parents' Cigarette Smoking to Outcome of Pregnancy—Implications as to the Problem of Inferring Causation from Observed Associations." As the title suggests, the paper is about the relationship between smoking during pregnancy, the weight of babies at birth, and mortality in the first month of life.

Based on data from about 13,000 babies born near San Francisco between 1960 and 1967, Yerushalmy reported:

- Babies of mothers who smoked were about 6% lighter at birth.
- Smokers were about twice as likely to have babies lighter than 2500 grams, which is considered "low birthweight."
- Low-birthweight babies were much more likely to die within a month of birth: the mortality rate was 174 per 1000 for low-birthweight babies and 7.8 per 1000 for others.

These results were not surprising. At that time, it was well known that children of smokers were lighter at birth and that low-birthweight babies were more likely to die.

Putting those results together, you might expect mortality rates to be higher for children of smokers. And you would be right, but the difference was not very big. For White mothers, the mortality rate was 11.3 per 1000 for children of smokers, compared to 11.0 for children of nonsmokers.

That's strange, but it gets even stranger. If we select only the low-birthweight (LBW) babies, we find:

- For LBW babies of nonsmokers, the mortality rate was 218 per 1000;
- For LBW babies of smokers, it was only 114 per 1000, about 48% lower.

Yerushalmy also compared rates of congenital anomalies (birth defects).

- For LBW babies of nonsmokers, the rate was 147 per 1000,
- For LBW babies of smokers, it was 72 per 1000, about 53% lower.

These results make maternal smoking seem *beneficial* for low-birthweight babies, somehow protecting them from birth defects and mortality. Yerushalmy concluded, "These paradoxical findings raise doubts and argue against the proposition that cigarette smoking acts as an exogenous factor which interferes with intrauterine development of the fetus." In other words, maybe maternal smoking isn't bad for babies after all.

Yerushalmy's paper was influential. In 2014 the *International Journal of Epidemiology* reprinted it along with an editorial retrospective and five invited commentaries. They report that "Yerushalmy's findings were widely publicized by tobacco companies during the 1970s, '80s and '90s." Media attention included headlines like "Mothers Needn't Worry, Smoking Little Risk to Baby" in 1971 and a column in *Family Health* magazine titled "In Defense of Smoking Moms" in 1972.

In 1973, after the US surgeon general reported a "strong, probably causal connection" between mothers' smoking and infant mortality, Yerushalmy sent a letter to the United States Senate asserting that the case for a causal connection had not been proved. In the United States, Yerushalmy's legacy might be responsible for "holding up anti-smoking measures among pregnant women for perhaps a decade," according to one commentary. Another suggests that in the UK, it "postponed by several years any campaign to change mothers' smoking habits."

But it was a mistake. At the risk of giving away the ending, the low-birthweight paradox is a statistical artifact. In fact, maternal smoking is harmful to babies, regardless of birthweight. It only seems beneficial because the analysis is misleading.

It took some time for the error to be discovered. In 1983, epidemiologists Allen Wilcox and Ian Russell published a partial explanation. Using computer simulations, they showed that if you have two groups with the same mortality rate and different average birthweights, you get a version of the birthweight paradox: the group with lower birthweights has more LBW babies, but they are healthier; that is, their mortality rate is lower, compared to LBW babies from the other group.

Their simulations show that the low-birthweight paradox can occur due to statistical bias, even if actual mortality rates are the same for both groups. But to me this conclusion is not entirely satisfying, partly because it is based on simulated data and partly because it does not explain *why* the paradox occurs.

A clearer explanation came in 2006 from epidemiologists at Harvard University and the National Institutes of Health (NIH), based on data from three million babies born in 1991. Using the same dataset, which is available from the National Center for Health Statistics (NCHS), I will replicate their results and summarize their explanation. Then I'll repeat the analysis with data from 2018, and we'll see what has changed.

THREE MILLION BABIES CAN'T BE WRONG

In the 1991 data from NCHS, about 18% of the mothers reported smoking during pregnancy, down from 37% in Yerushalmy's dataset from the 1960s. Babies of smokers were lighter on average than babies of nonsmokers by about 7%, which is comparable to the difference in the 1960s data. The following figure shows the distribution of weights for the two groups. The vertical line is at 2500 grams, the threshold for low birthweight.

Distributions of birthweights

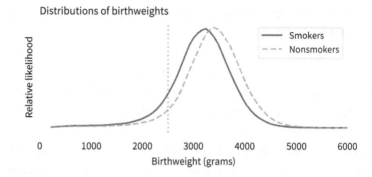

The shapes of the distributions are similar, but for smokers, the curve is shifted to the left. For mothers who smoked, the fraction of babies below 2500 grams is about 11%; for nonsmokers, it is only 6%. The ratio of these percentages is almost 2 to 1, about the same as in the 1960s. So the effect of smoking on birthweight has been consistent.

Overall infant mortality was substantially lower in 1991. In the 1960s dataset, about 13 per 1000 babies died within the first *month* of life; in 1991, about 8.5 per 1000 died in the first *year*. In 1991, the mortality rate was higher for babies of smokers, almost 12 per 1000, than for babies of nonsmokers, 7.7 per 1000. So the risk of mortality was 54% higher for babies of mothers who smoked.

In summary, babies of mothers who smoked were about twice as likely to be underweight compared to babies of nonsmokers, and underweight babies were about 50% more likely to die than normal-weight babies. However, if we select babies lighter than 2500 grams, the mortality rate is 20% *lower* for babies of smokers, compared to LBW babies of nonsmokers. The low-birthweight paradox strikes again.

The analysis so far is based on only two groups, babies born lighter or heavier than 2500 grams. But it might be a mistake to lump all LBW babies together. In reality, a baby born close to 2500 grams has a better chance of surviving than a baby born at 1500 grams. So, following the analysis in the 2006 paper, I partitioned the dataset into groups with similar birthweights and computed the mortality rate in each group. The following figure shows the results.

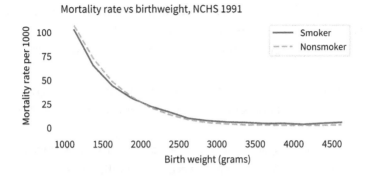

Mortality rate vs birthweight, NCHS 1991

This figure provides a more detailed view of the low-birthweight paradox. Among babies heavier than 2000 grams, mortality is higher for children of smokers, as expected. Among lighter babies, mortality is lower for children of smokers.

OTHER GROUPS

As it turns out, the low-birthweight paradox doesn't apply only to smokers and nonsmokers. The 2006 paper describe a similar effect for babies born at high altitude: they are lighter on average than babies born at low altitude, but if we select LBW babies, the mortality rate is lower for the ones born at high altitude.

And Yerushalmy reported another example. Babies of short mothers are lighter, on average, than babies of tall mothers. In his dataset, babies of short mothers were twice as likely to be LBW, but among LBW babies of short mothers, the mortality rate was 49% lower and the rate of birth defects was 34% lower.

Yerushalmy called the relationship between smokers vs. nonsmokers and short vs. tall mothers a "remarkable parallelism." But he did not recognize it as evidence that statistical bias is the explanation for both. Instead, he doubled down: "This comparison is presented not as proof that the differences between smokers and nonsmokers are necessarily of biological origin, rather it is to indicate that a biological hypothesis is not unreasonable." With the benefit of further research, we can see that Yerushalmy was mistaken. Smoking, high altitude, and short mothers do not protect low-birthweight babies from birth defects and mortality. Rather, they provide a relatively benign explanation for low birthweight.

To see why, suppose four things can cause low birthweight:

- The mother might be short, which is not at all harmful to the baby.
- The baby might be born at high altitude, which has little if any effect on mortality.
- The mother might be a smoker, which is somewhat harmful to the baby.
- The baby might have a birth defect, which greatly increases the rate of mortality.

Now suppose you are a doctor and you hear that a baby under your care was born underweight. You would be concerned, because you know that the baby faces a higher than average risk of mortality.

But suppose the baby was born in Santa Fe, New Mexico, at 2200 meters of elevation to a mother at only 150 cm of elevation (just under five feet). You would be relieved, because either of those factors might explain low birthweight, and neither implies a substantial increase in mortality. And if you learned that the mother was a smoker, that would be good news, too, because it provides another possible explanation for low birthweight, which means that the last and most harmful explanation is less likely. Maternal smoking is still bad for babies, but it is not as bad as birth defects.

It is frustrating that Yerushalmy did not discover this explanation. In retrospect, he had all the evidence he needed, including the smoking gun (sorry!): the rates of birth defects. We've seen that LBW babies of smokers are less likely to have birth defects, but that's not because maternal smoking somehow protects babies from congenital anomalies. It's because low birthweight generally has a cause, and if the cause is not smoking, it is more likely to be something else, including a birth defect. We can confirm that this explanation is correct by selecting babies with no congenital anomalies observed at birth. If we do that, we find that babies of smokers have higher mortality rates in nearly every weight category, as expected.

Maternal smoking might be less harmful than many birth defects, but to be clear, it is still harmful. And Yerushalmy's error might be understandable, but it was also harmful. At a time when the health risks of smoking were still contested, his paper created confusion and gave cover to people with an interest in minimizing the dangers.

THE END OF THE PARADOX

When a paradoxical phenomenon is explained, it ceases to be a paradox. And in the case of the low-birthweight paradox, at about the same time it also ceased to be a phenomenon. In the most recent NCHS dataset, including 3.8 million babies born in 2018, the low-birthweight paradox has disappeared. In this dataset, only 6% of the mothers reported smoking during pregnancy, down from 18% in 1991 and 37% in the 1960s. Babies of smokers were lighter on average than babies of nonsmokers by about 6%, comparable to the difference in the previous two datasets. So the effect of smoking on birthweight has been constant for 60 years.

In 2018, fewer babies died in the first year of life; the mortality rate was 5.5 per 1000, down from 8.5 in 1991. And the mortality rate for babies of smokers was more than twice the rate for babies of nonsmokers, almost 11 per 1000 compared to 5.1. Again, we can partition the dataset into groups with similar birthweights and compute the mortality rate in each group. The following figure shows the results.

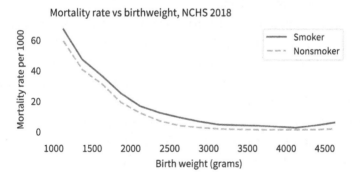

At every birthweight, mortality is higher for children of smokers. For better or worse, the low-birthweight paradox is no more.

A TWIN PARADOX

The low-birthweight paradox might seem like an esoteric issue, but similar phenomena are an ongoing source of confusion. As an example, in 2000, researchers at the University of Hong Kong reported a study of 1.5 million babies born in Sweden between 1982 and 1995. They compute mortality rates for single births, twins, and triplets as a function of gestational age (that is, number of weeks of pregnancy) and find that

- twins and triplets are more likely to be born preterm—that is, prior to 37 weeks of gestational age;
- babies born preterm have higher mortality rates than babies born full-term; and
- overall mortality rates are higher for twins and triplets.

These results are unsurprising. However, when they select babies born preterm, they find that survival rates are higher for twins and triplets, compared to single births.

To explain this surprising result, they suggest "twins have better health than singletons initially"—before 36 weeks of gestation—but "they could not enjoy the benefit of a longer gestational duration as much as singletons could." That is, they suggest a biological explanation, not a statistical one.

In an invited commentary on their paper, medical statistician Rolv Lie reports similar results from 1.8 million babies born in Norway between 1967 and 1998. However, he comes to a different conclusion, one that I hope has occurred to you after reading this far. He suggests that the difference in mortality between single and multiple births "does not reflect the frailty among undelivered fetuses at the same gestational ages." Rather, "Preterm babies have more than the prematurity problem of their specific gestational age, for they also suffer from the pathological causes of their preterm delivery, which are usually unknown."

In less flowery language, multiple pregnancy is one of several things that can cause preterm birth, and among them, it is relatively harmless. So if a preterm baby is a twin or triplet, it is less likely that they suffer from some other, more harmful, condition.

THE OBESITY PARADOX

The low-birthweight paradox is no more, and the twin paradox has been explained, but the obesity paradox is alive and well.

The first example of the obesity paradox was reported in 1999. Although it was well known that obesity is a risk factor for kidney disease and that kidney disease is often fatal, researchers found that among patients undergoing dialysis due to kidney failure, survival times for obese patients were *longer* than for other patients. They suggested as a possible explanation: "Overweight patients have an increase in adipose tissue and, therefore, are less likely to suffer from energy deficits." Based on this conclusion, they recommended that "proper nutrition to maintain a high-end normal BMI [body mass index] should help reduce the high mortality and morbidity rates" in patients on dialysis.

Since then, similar patterns have been reported for *many* other diseases. Among them:

- Obesity increases the risk of stroke, myocardial infarction, heart failure, and diabetes.
- All of these conditions are associated with increased mortality and morbidity.
- Nevertheless, if we select patients with any of these conditions, we find that obese patients have *lower* mortality and morbidity than normal-weight patients.

To get a sense of the confusion these results have created, you only have to read the titles of the papers written about them. Here's a sample from a quick search:

- 2006: "The Obesity Paradox: Fact or Fiction?"
- 2007: "Obesity-Survival Paradox: Still a Controversy?"
- 2010: "The Obesity Paradox: Perception vs. Knowledge"

- 2011: "Effect of Body Mass Index on Outcomes after Cardiac Surgery: Is There an Obesity Paradox?"
- 2013: "Obesity Paradox Does Exist"
- 2019: "Obesity Paradox in Cardiovascular Disease: Where Do We Stand?"

So, where *do* we stand? In my opinion, the most likely explanation is the one offered in a letter to the editor of *Epidemiology* with the promising title, "The 'Obesity Paradox' Explained." As an example, the authors consider the obesity paradox in patients with heart failure and propose the following explanation:

- Obesity is one cause of heart failure, and heart failure causes mortality.
- But there are other causes of heart failure, which also cause mortality.
- Compared to other causes, obesity is relatively benign.
- If a patient with heart failure is obese, the other causes are less likely.
- Therefore, among patients with heart failure, obese patients have lower mortality.

Using data from the National Health and Nutrition Examination Survey and the National Death Index, they show that this explanation is plausible; that is, the statistical relationships implied by their theory appear in the data, and they are strong enough to explain the obesity paradox, even if obesity has no protective effect.

I think their argument is convincing, but it does not rule out other causal mechanisms; for example, in the case of kidney failure, fat tissue might prevent organ damage by diluting the toxins that accumulate when kidney function is diminished. And it does not rule out other statistical explanations; for example, if long-term severe disease causes weight loss, obesity might be an indicator of relatively short-term mild disease. Nevertheless, until there is evidence for other explanations, the title of my article is "The Obesity Paradox: No."

BERKSON'S TOASTER

At some point while you were reading this chapter, you might have noticed connections between the low-birthweight paradox and Berkson's paradox.

With Berkson's paradox, if there are two ways to be included in a sample, we often find that the alternatives are negatively correlated in the sample, even if they are unrelated, or positively correlated, in the general population. For example, to dunk a basketball, you have to be tall or you have to be able to jump. If we sample people who can dunk, we find that the shorter people can jump higher, even if there is no such relationship in general.

With the low-birthweight paradox, again, there are two causes for the same effect; if we select for the effect, we often find that the alternatives are negatively correlated in the sample. And if one of the causes is more harmful, the other might seem relatively benign. For example, if jumping is bad for your knees, we might find that, among people who can dunk, taller players have healthier knees. (I think I just invented a new paradox.)

When you think of the low-birthweight paradox, and the related paradoxes in this chapter, I suggest you remember what I call "Berkson's toaster." Suppose you hear a smoke alarm in your kitchen. You get up and move quickly to the kitchen, where you find that someone has left a piece of toast in the toaster too long. You would probably feel relieved. Why? Because of all the things that could cause a smoke alarm, burnt toast is probably the least harmful. That doesn't mean burning toast is good, but if the alarm is sounding, burnt toast is better than the alternatives.

CAUSAL DIAGRAMS

The 2006 paper explaining the low-birthweight paradox and the 2013 paper explaining the obesity paradox are noteworthy because they use causal diagrams to represent hypothetical causes and their effects. For example, here is a causal diagram that represents an explanation for the low-birthweight paradox:

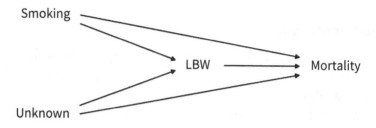

Each arrow represents a causal relationship, so this diagram represents the following hypotheses:

- Maternal smoking causes low birthweight (LBW) and mortality, in the sense that it increases the probability of both.
- Additional unknown factors, including birth defects, also cause both LBW and mortality.
- LBW also causes mortality (regardless of what caused it).

Importantly, these arrows represent truly causal relationships, not just statistical association. For example, the arrow from LBW to mortality means that low birthweight is harmful in and of itself, not only because it is statistical evidence of a harmful condition. And there is no arrow from mortality to LBW because, even though they are correlated, mortality does not cause low birthweight.

However, this diagram does not contain all of the information we need to explain the low-birthweight paradox, because it does not represent the strengths of the different causal relationships. In order to explain the paradox, the total contribution of the unknown factors to mortality has to exceed the total contribution of maternal smoking. Also, this diagram is not the only one that could explain the low-birthweight paradox. For example, the arrow from LBW to mortality is not necessary; the paradox could happen even if low birthweight were entirely harmless. Nevertheless, diagrams like this are a useful way to document a hypothetical set of causal relationships.

The following causal diagram represents the explanation of the obesity paradox proposed in the 2013 paper.

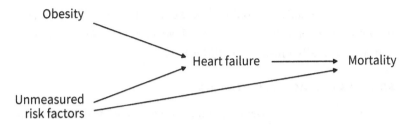

The arrows in this diagram represent the following hypotheses:

- Obesity causes heart failure, in the sense that it increases its probability.
- Other risk factors, including "genetic factors and lifestyle behaviors" cause both heart failure and mortality.
- Heart failure causes mortality.

In this model, obesity causes mortality indirectly, by increasing the probability of heart failure, but it is not a direct cause of mortality. This assumption implies that an obese patient who does not have heart failure would not have increased mortality. In reality, that's probably not true; for example, obesity also causes diabetes, which increases mortality. But we can leave that out of the model because it is not necessary to explain the paradox. What is necessary, but not shown in the diagram, is that the total contribution of the unknown factors to mortality must exceed the contribution of obesity indirectly through heart failure.

Comparing these two diagrams, we can see what the two paradoxes have in common:

- In the low-birthweight paradox, the condition we selected has two causes, maternal smoking and other factors like birth defects.
- In the obesity paradox, the condition we selected has two causes, obesity and other genetic and lifestyle factors.

Likewise, when we select babies born preterm, that condition has two causes, multiple birth and other risk factors.

In the context of causal modeling, a condition that has two (or more) causes is called a "collider," because in the causal diagram, the incoming arrows collide. So what I've been calling Berkson's paradox is more generally known as collider bias.

SOURCES AND RELATED READING

- Yerushalmy's paper was originally published in the *American Journal of Epidemiology* [133] and reprinted in the *International Journal of Epidemiology* [134].
- The editorial introduction to the reprinted version is by Shah Ebrahim [37], with commentaries by Keyes, Davey Smith, and Susser [59]; Parascandola [91]; Goldstein [48]; Kramer, Zhang, and Platt [62]; and VanderWeele [128].
- The paper that demonstrated the low-birthweight paradox with a computer simulation is by Wilcox and Russell [131].
- The paper that demonstrated it with NCHS data is by Hernández-Díaz, Schisterman, and Hernán [52].
- The NCHS data is available from the Centers for Disease Control and Prevention (CDC) [130].
- The paper reporting the twin paradox is by Cheung, Yip, and Karlberg [17]; the commentary explaining it is by Rolv Lie [68].
- The paper reporting the obesity paradox is by Fleischmann et al. [42]; the 2013 paper explaining it is by Banack and Kaufman [10].
- If you would like to read more about causal modeling, a good introduction for a general audience is Judea Pearl's *The Book of Why* [93].

CHAPTER 8

THE LONG TAIL OF DISASTER

You would think we'd be better prepared for disaster. But events like Hurricane Katrina in 2005, which caused catastrophic flooding in New Orleans, and Hurricane Maria in 2017, which caused damage in Puerto Rico that has still not been repaired, show that large-scale disaster response is often inadequate. Even wealthy countries—with large government agencies that respond to emergencies and well-funded organizations that provide disaster relief—have been caught unprepared time and again.

The are many reasons for these failures, but one of them is that rare, large events are fundamentally hard to comprehend. Because they are rare, it is hard to get the data we need to estimate precisely how rare. And because they are large, they challenge our ability to imagine quantities that are orders of magnitude bigger than what we experience in ordinary life. My goal in this chapter is to present the tools we need to comprehend the small probabilities of large events so that, maybe, we will be better prepared next time.

THE DISTRIBUTION OF DISASTER

Natural and human-caused disasters are highly variable. The worst disasters, in terms of both lives lost and property destroyed, are thousands of times bigger than smaller, more common disasters. That fact might not be surprising, but there is a pattern to these costs that is less obvious. One way to see this pattern is to plot the magnitudes of the biggest disasters on a logarithmic scale. To demonstrate, I'll use the estimated costs from a list of 125 disasters on Wikipedia.

Most of the disasters on the list are natural, including 56 tropical cyclones and other windstorms, 16 earthquakes, 8 wildfires, 8 floods, and 6 tornadoes. The costliest natural disaster on the list is the 2011 Tōhoku earthquake and tsunami, which caused the Fukushima nuclear disaster, with a total estimated cost of $424 billion in 2021 dollars. There are not as many human-made disasters on the list, but the costliest is one of them: the Chernobyl disaster, with an estimated cost of $775 billion. The list of human-made disasters includes four other instances of contamination with oil or radioactive material, three space flight accidents, two structural failures, two explosions, and two terror attacks.

This list is biased: disasters that affect wealthy countries are more likely to damage high-value assets, and the reckoning of the costs is likely to be more complete. A disaster that affects more lives—and more severely—in other parts of the world might not appear on the list. Nevertheless, it provides a way to quantify the sizes of some large disasters. The following figure shows a "rank-size plot" of the data: the x-axis shows the ranks of the disasters from 1 to 125, where 1 is the most costly; the y-axis shows their costs in billions of US dollars.

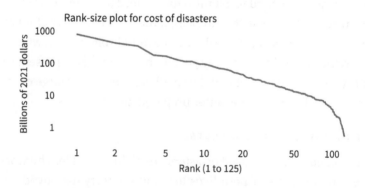

With both axes on a log scale, the rank-size plot is a nearly straight line, at least for the top 100 disasters. And the slope of this line is close to 1, which implies that each time we double the rank, we cut the costs by one-half. That is, compared to the costliest disaster, the cost of the second-worst is one-half; the cost of the fourth-worst is one-quarter; the cost of the eighth is one-eighth, and so on.

In this chapter I will explain where this pattern comes from, but I'll warn you that the answers we have are incomplete. There are several ways that natural and engineered systems can generate distributions like this. In some cases, like impact craters and the asteroids that form them, we have physical models that explain the distribution of their sizes. In other cases, like solar flares and earthquakes, we only have hints.

We will also use this pattern to make predictions. To do that, we have to understand "long-tailed" distributions. In chapter 1 we saw that many measurements, like human height, follow a Gaussian distribution. And in chapter 4 we saw that many other measurements, like human weight, follow a lognormal distribution. I suggested a general explanation for these patterns:

- If the things we measure are the result of adding many factors, the sum tends to be Gaussian.
- If the things we measure are the result of multiplying many factors, the product tends to be lognormal.

The difference between these distributions is most noticeable in the tails—that is, in the distribution of the most extreme values. Gaussian distributions are symmetric, and their tails don't extend far from the mean. Lognormal distributions are asymmetric; usually the tail extends farther to the right than the left, and it extends farther from the mean.

If you think measurements follow a Gaussian distribution, and they actually follow a lognormal distribution, you will be surprised to find outliers much farther from the mean than you expect. That's why the fastest runners are much faster than ordinary runners and the best chess players are so much better than casual players. For this reason, the lognormal distribution is considered a "long-tailed" distribution. But it is not the only long-tailed distribution, and some of the others have tails even longer than lognormal. So, if you think measurements follow a lognormal distribution, and they actually follow one of these "longer than lognormal" distributions, you will be surprised to find outliers much, much farther from the mean than you expect.

Let's get back to the distribution of disaster sizes, and we'll see the difference one of these long-tailed distributions can make. To visualize disaster sizes, I will show the "tail distribution," which is the complement of the cumulative distribution function (CDF). In chapter 5, we used a tail distribution to represent survival times for cancer patients. In that context, it shows the fraction of patients that survive a given time after diagnosis. In the context of disasters, it shows the fraction of disasters whose costs exceed a given threshold. The following figure shows the tail distribution of the 125 costliest disasters on a log scale, along with a lognormal model.

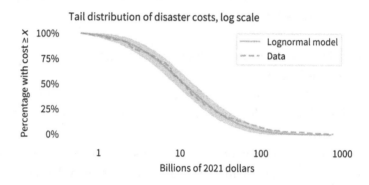

The dashed line shows the fraction of disasters that exceed each magnitude of cost. For example, 98% of the disasters exceed a cost of $1 billion, 54% exceed $10 billion, and 7% exceed $100 billion. The solid line shows a lognormal model I chose to fit the data; the shaded area shows how much variation we expect with a sample size of 125. Most of the data fall within the bounds of the shaded area, which means that they are consistent with the lognormal model.

However, if you look closely, there are more large-cost disasters than we expect in a lognormal distribution. To see this part of the distribution more clearly, we can plot the y-axis on a log scale; the following figure shows what that looks like.

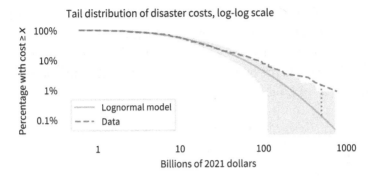

Tail distribution of disaster costs, log-log scale

Again, the shaded area shows how much variation we expect in a sample of this size from a lognormal model. In the extreme tail of the distribution, this area is wide because there are only a few observations in this range: where we have less data, we have more uncertainty.

This view of the distribution is like a microscope that magnifies the tail. Now we can see more clearly where the model diverges from the data and by how much. For example, the vertical dotted line shows the discrepancy at $500 billion. According to the model, only one out of 1000 disasters should exceed this cost, but the actual rate is 16 per 1000. If your job is to prepare for disasters, an error that large could be, well, disastrous.

To make better predictions, we need a model that accurately estimates the probability of large disasters. In the literature of long-tailed distributions, there are several to choose from; the one I found that best matches the data is Student's *t* distribution. It was described in 1908 by William Sealy Gosset, using the pseudonym "Student," while he was working on statistical problems related to quality control at the Guinness Brewery.

The shape of the *t* distribution is a bell curve similar to a Gaussian, but the tails extend farther to the right and left. It has two parameters that determine the center point and width of the curve, like the Gaussian distribution, plus a third parameter that controls the thickness of the tails, that is, what fraction of the values extend far from the mean.

Since I am using a *t* distribution to fit the logarithms of the values, I'll call the result a log-*t* model. The following figure shows the distribution of costs again, along with a log-*t* model I chose to fit the data. In the top panel, the *y*-axis shows the fraction of disasters on a linear scale; in the bottom panel, the *y*-axis shows the same fractions on a log scale.

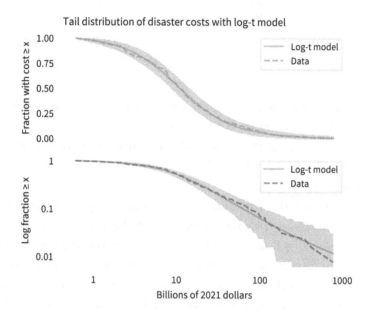

In the top panel, we see that the data fall within the shaded area over the entire range of the distribution. In the bottom panel, we see that the model fits the data well even in the extreme tail. The agreement between the model and the data provides a hint about why the distribution might have this shape. Mathematically, Student's *t* distribution is a mixture of Gaussian distributions with different standard deviations. In fact, the list of disasters includes several different kinds, which might have different distributions of cost, so this explanation is plausible. To see whether it holds up, let's take a closer at just one cause of disaster: earthquakes.

EARTHQUAKES

To describe the magnitudes of earthquakes, I downloaded data from the Southern California Earthquake Data Center. Their archive goes all the way back to 1932, but the sensors they use and their coverage have changed over that time. For consistency, I selected data from January 1981 to April 2022, which includes records of 791,329 earthquakes.

The magnitude of an earthquake is measured on the "moment magnitude scale," which is an updated version of the more well-known Richter scale. A moment magnitude is proportional to the logarithm of the energy released by an earthquake. A two-unit difference on this scale corresponds to a factor of 1000, so a magnitude-5 earthquake releases 1000 times more energy than a magnitude-3 earthquake and 1000 times less energy than a magnitude-7 earthquake.

The following figure shows the distribution of moment magnitudes compared to a Gaussian model, which is a lognormal model in terms of energy.

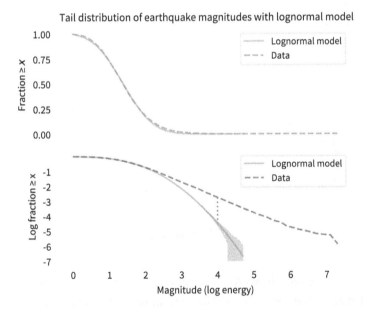

If you only look at the top panel, you might think the lognormal model is good enough. But the probabilities for magnitudes greater than 3 are so small that differences between the model and the data are not visible.

The bottom panel, with the y-axis on a log scale, shows the tail more clearly. Here we can see that the lognormal model drops off more quickly than the data. According to the model, the fraction of earthquakes with magnitude 4 or more should be about 33 per million. But the actual fraction of earthquakes this big is about 1800 per million, more than 50 times higher.

For larger earthquakes, the discrepancy is even bigger. According to the model, fewer than one earthquake in a million should exceed magnitude 5; in reality, about 170 per million do. If your job is to predict large earthquakes, and you underestimate their frequency by a factor of 170, you are in trouble. The following figure shows the same distribution compared to a log-t model.

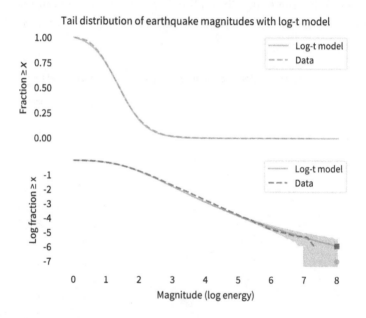

The top panel shows that the log-t model fits the distribution well for magnitudes less than 3. The bottom panel shows that it also fits the tail of the distribution.

If we extrapolate beyond the data, we can use this model to compute the probability of earthquakes larger than the ones we have seen so far. For example, the square marker shows the predicted probability of an earthquake with magnitude 8, which is about 1.2 per million.

The number of earthquakes in this dataset is about 18,800 per year; at that rate we should expect an earthquake that exceeds magnitude 8 every 43 years, on average. To see how credible that expectation is, let's compare it to predictions from sources more knowledgeable than me. In 2015, the US Geological Survey (USGS) published the third revision of the Uniform California Earthquake Rupture Forecast (UCERF3). It predicts that we should expect earthquakes in Southern California that exceed magnitude 8 at a rate of one per 522 years. The circle marker in the figure shows the corresponding probability.

Their forecast is compatible with mine in the sense that it falls within the shaded area that represents the uncertainty of the model. In other words, my model concedes that the predicted probability of a magnitude-8 earthquake could be off by a factor of 10, based on the data we have. Nevertheless, if you own a building in California, or provide insurance for one, a factor of 10 is a big difference. So you might wonder which model to believe, theirs or mine.

To find out, let's see how their predictions from 2015 have turned out so far. For comparison, I ran the log-t model again using only data from prior to 2015. The following figure shows their predictions, mine, and the actual number of earthquakes that exceeded each magnitude in the 7.3 years between January 2015 and May 2022.

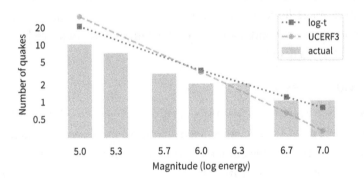

Starting on the left, the UCERF model predicted that there would be 30 earthquakes in this interval with magnitude 5.0 or more. The log-t model predicted 20. In fact, there were 10. So there were fewer large earthquakes than either model predicted. Near the middle of the figure, both models predicted three earthquakes with magnitude 6.0 or more. In fact, there were two, so both models were close. On the right, the UCERF model expects 0.3 earthquakes with magnitude 7.0 or more in 7.3 years. The log-t expects 0.8 earthquakes of this size in the same interval. In fact, there was one: the main shock of the 2019 Ridgecrest earthquake sequence measured 7.1.

The discrepancy between the two models increases for larger earthquakes. The UCERF model estimates 87 years between earthquakes that exceed 7.5; the log-t model estimates only 19. The UCERF model estimates 522 years between earthquakes that exceed 8.0; the log-t model estimates only 36. Again, you might wonder which model to believe.

According to their report, UCERF3 was "developed and reviewed by dozens of leading scientific experts from the fields of seismology, geology, geodesy, paleoseismology, earthquake physics, and earthquake engineering. As such, it represents the best available science with respect to authoritative estimates of the magnitude, location, and likelihood of potentially damaging earthquakes throughout the state."

One the other hand, my model was developed by a lone data scientist with no expertise in any of those fields. And it is a purely statistical model; it is based on the assumption that the processes that generate earthquakes are obliged to follow simple mathematical models, even beyond the range of the data we have observed.

I'll let you decide.

SOLAR FLARES

While we wait for the next big earthquake, let's consider another source of potential disasters: solar flares. A solar flare is an eruption of light and other electromagnetic radiation from the atmosphere of the Sun that can last from several seconds to several hours. When this radiation reaches Earth, most of it is absorbed by our atmosphere, so it usually has little effect at the surface. Large solar flares can interfere with radio communication and GPS, and they are dangerous to astronauts, but other than that they are not a serious cause for concern.

However, the same disruptions that cause solar flares also cause coronal mass ejections (CME) which are large masses of plasma that leave the Sun and move into interplanetary space. They are more narrowly directed than solar flares, so not all of them collide with a planet, but if they do, the effects can be dramatic. As a CME approaches Earth, most of its particles are steered away from the surface by our magnetic field. The ones that enter our atmosphere near the poles are sometimes visible as auroras.

However, very large CMEs cause large-scale, rapid changes in the Earth's magnetic field. Before the Industrial Revolution, these changes might have gone unnoticed, but in the past few hundred years, we have crisscrossed the surface of the Earth with millions of kilometers of wires. Changes in the Earth's magnetic field with the magnitude created by a CME can drive large currents through these wires. In 1859, a CME induced enough current in telegraph lines to start fires and shock some telegraph operators. In 1989, a CME caused a large-scale blackout in the electrical grid in Quebec, Canada.

If we are unlucky, a very large CME could cause widespread damage in electrical grids that could take years and trillions of dollars to repair. Fortunately, these risks are manageable. With satellites positioned between the Earth and Sun, we have the ability to detect an incoming CME hours before it reaches Earth. And by shutting down electrical grids for a short time, we should be able to avoid most of the potential damage.

So, on behalf of civilization, we have a strong interest in understanding solar flares and coronal mass ejections. Toward that end, the Space Weather Prediction Center (SWPC) monitors and predicts solar activity and its effects on Earth. Since 1974 it has operated the Geostationary Operational Environmental Satellite system (GOES). Several of the satellites in this system carry sensors that measure solar flares. Data from these sensors, going back to 1975, is available for download.

Since 1997, these datasets include "integrated flux"—the kind of term science fiction writers love—which measures the total energy from a solar flare that would pass through a given area. The magnitude of this flux is one way to quantify its potential for impact on Earth. This dataset includes measurements from more than 36,000 solar flares. The following figure shows the distribution of their magnitudes compared to a lognormal model.

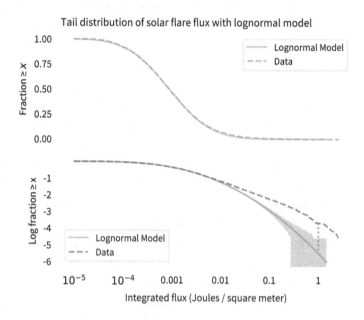

The x-axis shows integrated flux on a log scale; the largest flare observed during this period measured 2.6 Joules per square meter on September 7, 2005. By itself, that number might not mean much, but it is almost a million times bigger than the smallest flare. So the thing to appreciate here is the great range of magnitudes.

Looking at the top panel, it seems like the lognormal model fits the data well. But for fluxes greater than 0.1 Joules per square meter, the probabilities are so small they are indistinguishable. Looking at the bottom panel, we can see that for fluxes in this range, the lognormal model underestimates the probabilities. For example, according to the model, the fraction of flares with flux greater than 1 should be about three per million; in the dataset, it is about 200 per million. The dotted line in the figure shows this difference. If we are only interested in flares with flux less than 0.01, the lognormal model might be good enough. But for predicting the effect of space weather on Earth, it is the largest flares we care about, and for those, the lognormal model might be dangerously inaccurate.

Let's see if the log-*t* model does any better. The following figure shows the distribution of fluxes again, along with a log-*t* distribution.

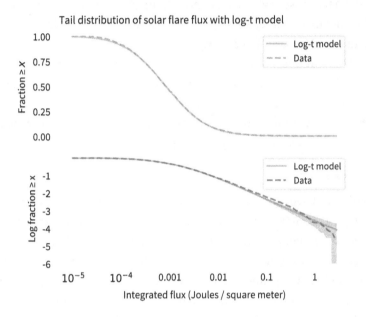

The top panel shows that the log-*t* model fits the data well over the range of the distribution, except for some discrepancy among the smallest flares. Maybe we don't detect small flares as reliably, or maybe there are just fewer of them than the model predicts. The bottom panel shows that the model fits the tail of the distribution except possibly for the most extreme values. Above 1 Joule per meter squared, the actual curve drops off faster than the model, although it is still within the range of variation we expect. There might be a physical reason for this drop-off; that is, we might be approaching the maximum flux a solar flare on our Sun can produce. But there are only seven flares in the dataset with fluxes greater than 1, so it may just be lucky that they are a little smaller than expected.

This example shows that the distribution of the size of solar flares is longer-tailed than a lognormal distribution, which suggests that the largest possible solar flares might be substantially larger than the ones we have seen so far. If we use the model to extrapolate beyond the data—which is always an uncertain thing to do—we expect 2.4 flares per million to exceed 10 Joules per square meter, and 0.36 per million to exceed 100. In the past 20 years we have observed about 36,000 flares, so it would take more than 500 years to observe a million. But in that time, we might see a solar flare 10 or 100 times bigger than the biggest we have seen so far.

Observations from the Kepler space telescope have found that a small fraction of stars like our Sun produce "superflares" up to 10,000 times bigger than the ones we have observed. At this point we don't know whether these stars are different from our Sun in some way, or whether the Sun is also capable of producing superflares.

LUNAR CRATERS

While we have our eyes on the sky, let's consider another source of potential disaster: an asteroid strike. On March 11, 2022, an astronomer near Budapest, Hungary detected a new asteroid, now named 2022 EB5, on a collision course with Earth. Less than two hours later, it exploded in the atmosphere near Greenland. Fortunately, no large fragments reached the surface, and they caused no damage.

We have not always been so lucky. In 1908, a much larger asteroid entered the atmosphere over Siberia, causing an estimated two-megaton explosion, about the same size as the largest thermonuclear device tested by the United States. The explosion flattened something like 80 million trees in an area covering 2100 square kilometers, almost the size of Rhode Island. Fortunately, the area was almost unpopulated; a similar-sized impact could destroy a large city.

These events suggest that we would be wise to understand the risks we face from large asteroids. To do that, we'll look at evidence of damage they have done in the past: the impact craters on the Moon. The largest crater on the near side of the moon, named Bailly, is 303 kilometers in diameter; the largest on the far side, the South

Pole-Aitken basin, is roughly 2500 kilometers in diameter. In addition to large, visible craters like these, there are innumerable smaller craters. The Lunar Crater Database catalogs nearly all of the ones larger than one kilometer, about 1.3 million in total. It is based on images taken by the Lunar Reconnaissance Orbiter, which NASA sent to the Moon in 2009. The following figure shows the distribution of these crater sizes, compared to a log-t model.

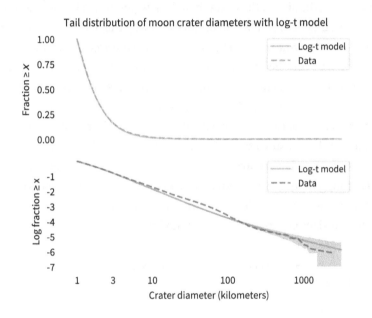

Since the dataset does not include craters smaller than one kilometer, I cut off the model at the same threshold. We can assume that there are many smaller craters, but with this dataset, we can't tell what the distribution of their sizes looks like. The log-t model fits the distribution well, but it's not perfect: there are more craters near 100 km than the model expects, and fewer craters larger than 1000 km. As usual, the world is under no obligation to follow simple rules, but this model does pretty well.

We might wonder why. To explain the distribution of crater sizes, it helps to think about where they come from. Most of the craters on the Moon were formed during a period in the life of the solar system called the "Late Heavy Bombardment," about four billion years ago. During this period, an unusual number of asteroids were displaced from the asteroid belt—possibly by interactions with large outer planets—and some of them collided with the Moon.

ASTEROIDS

As you might expect, there is a relationship between the size of an asteroid and the size of the crater it makes: in general, a bigger asteroid makes a bigger crater. So, to understand why the distribution of crater sizes is long-tailed, let's consider the distribution of asteroid sizes.

The Jet Propulsion Laboratory (JPL) and NASA provide data related to asteroids, comets and other small bodies in our solar system. From their Small-Body Database, I selected asteroids in the "main asteroid belt" between the orbits of Mars and Jupiter. There are more than a million asteroids in this dataset, about 136,000 with known diameter. The largest are Ceres (940 kilometers in diameter), Vesta (525 km), Pallas (513 km), and Hygeia (407 km). The smallest are less than one kilometer. The following figure shows the distribution of asteroid sizes compared to a log-*t* model.

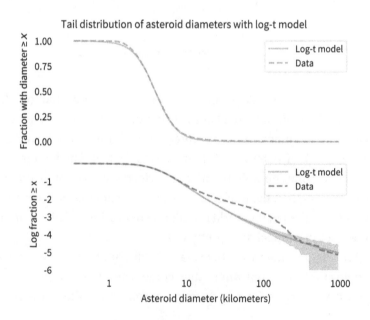

In the top panel, we see that the log-t model fits the data well in the middle of the range, except possibly near one kilometer. In the bottom panel, we see that the log-t model does not fit the tail of the distribution particularly well: there are more asteroids near 100 km than the model predicts. So the distribution of asteroid sizes does not strictly follow a log-t distribution. Nevertheless, we can use it to explain the distribution of crater sizes, as I'll show in the next section.

ORIGINS OF LONG-TAILED DISTRIBUTIONS

One of the reasons long-tailed distributions are common in natural systems is that they are persistent; for example, if a quantity comes from a long-tailed distribution and you multiply it by a constant or raise it to a power, the result follows a long-tailed distribution.

Long-tailed distributions also persist when they interact with other distributions. When you add together two quantities, if either comes from a long-tailed distribution, the sum follows a long-tailed distribution, regardless of what the other distribution looks like. Similarly, when you multiply two quantities, if either comes from a long-tailed distribution, the product usually follows a long-tailed distribution. This property might explain why the distribution of crater sizes is long-tailed.

Empirically—that is, based on data rather than a physical model—the diameter of an impact crater depends on the diameter of the projectile that created it (raised to the power 0.78) and on the impact velocity (raised to the power 0.44). It also depends on the density of the asteroid and the angle of impact. As a simple model of this relationship, I'll simulate the crater formation process by drawing asteroid diameters from the distribution in the previous section and drawing the other factors—density, velocity, and angle—from a lognormal distribution with parameters chosen to match the data. The following figure shows the results from this simulation along with the actual distribution of crater sizes.

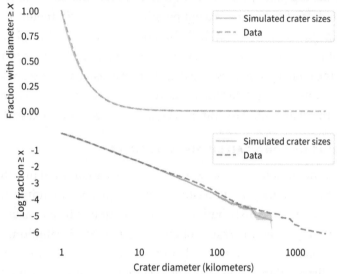

Tail distribution of moon crater diameters with simulation results

The simulation results are a good match for the data. This example suggests that the distribution of crater sizes can be explained by the relationship between the distributions of asteroid sizes and other factors like velocity and density.

In turn, there are physical models that might explain the distribution of asteroid sizes. Our best current understanding is that the asteroids in the asteroid belt were formed by dust particles that collided and stuck together. This process is called "accretion," and simple models of accretion processes can yield long-tailed distributions. So it may be that craters are long-tailed because of asteroids, and asteroids are long-tailed because of accretion.

In *The Fractal Geometry of Nature*, Benoit Mandelbrot proposes what he calls a "heretical" explanation for the prevalence of long-tailed distributions in natural systems: There may be only a few systems that generate long-tailed distributions, but interactions between systems might cause them to propagate. He suggests that data we observe are often "the joint effect of a fixed underlying true distribution and a highly variable filter," and "a wide variety of filters leave their asymptotic behavior unchanged."

The long tail in the distribution of asteroid sizes is an example of "asymptotic behavior." And the relationship between the size of an asteroid and the size of the crater it makes is an example of a "filter." In this relationship, the size of the asteroid gets raised to a power and multiplied by a "highly variable" lognormal distribution. These operations change the location and spread of the distribution, but they don't change the shape of the tail. When Mandelbrot wrote in the 1970s, this explanation might have been heretical, but now long-tailed distributions are more widely known and better understood. What was heresy then is orthodoxy now.

STOCK MARKET CRASHES

Natural disasters cost lives and destroy property. By comparison, stock market crashes might seem less frightening, but while they don't cause fatalities, they can destroy more wealth, more quickly. For example, the stock market crash on October 19, 1987, known as Black Monday, caused total worldwide losses of $1.7 trillion. The amount of wealth that disappeared in one day, at least on paper, exceeded the total cost of every earthquake in the twentieth century and so far in the twenty-first.

Like natural disasters, the distribution of magnitudes of stock market crashes is long-tailed. To demonstrate, I'll use data from the MeasuringWorth Foundation which has compiled the value of the Dow Jones Industrial Average at the end of each day from February 16, 1885 to the present, with adjustments at several points to make the values consistent. The series I collected ends on May 6, 2022, so it includes 37,512 days.

In percentage terms, the largest single-day drop was 22.6%, during the Black Monday crash. The second largest drop was 12.9% on March 16, 2020, during the crash caused by the COVID-19 pandemic. Numbers three and four on the list were 12.8% and 11.7%, on consecutive days during the Wall Street crash of 1929.

On the positive side, the largest single-day gain was on March 15, 1933, when the index gained 15.3%. There were other large gains in October 1929 and October 1931. The largest gain of the twenty-first century, so far, was on March 24, 2020, when the index gained 11.4% in expectation that the US Congress would pass a large economic stimulus bill.

Since we are interested in stock market crashes, I selected the 17,680 days when the value of the index dropped. The following figure shows the distribution of these negative percentage changes compared to a log-*t* model.

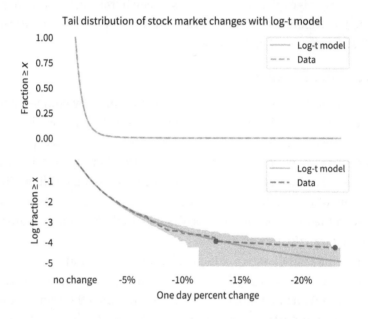

The top panel shows that the log-*t* model fits the left side of the distribution well. The bottom panel shows that it fits the tail well, with the possible exception of the last data point, the 1987 crash, which falls just at the edge of variation we expect. According to the model, there is only a 5% chance that we would see a crash as big as this one after 17,680 down days.

There are two interpretations of this observation: maybe Black Monday was a deviation from the model, or maybe we were just unlucky. These possibilities are the topic of some controversy, as I'll discuss in the next section.

BLACK AND GRAY SWANS

If you have read *The Black Swan*, by Nassim Taleb, the examples in this chapter provide context for understanding black swans and their relatives, gray swans. In Taleb's vocabulary, a black swan is a large, impactful event that was considered extremely unlikely before it happened, based on a model of prior events. If the distribution of event sizes is actually long-tailed, and the model is Gaussian, black swans will happen with some regularity.

For example, if we use a Gaussian distribution to model the magnitudes of earthquakes, we predict that the probability of a magnitude-7 earthquake is about four per sextillion (10^{18}). The actual rate in the dataset we looked at is about six per million, which is more likely by a factor of a trillion (10^{12}). When a magnitude-7 earthquake occurs, it would be a black swan if your predictions were based on the Gaussian model.

However, black swans can be "tamed" by using appropriate long-tailed distributions, as we did in this chapter. According to the log-t model, based on the distribution of past earthquakes, and the UCERF3 model, which is based on geological physics in addition to data, we expect a magnitude-7 quake to occur once every few decades, on average. To someone whose predictions are based on these models, such an earthquake would not be particularly surprising. In Taleb's vocabulary, it would be a gray swan: large and impactful, but not unexpected. This part of Taleb's thesis is uncontroversial: if you use a model that does not fit the data, your predictions will probably fail. In this chapter, we've seen several examples that demonstrate this point.

However, Taleb goes farther, asserting that long-tailed distributions "account for a few black swans, but not all." He explains, "A gray swan concerns modelable extreme events, a black swan is about unknown unknowns." A superflare might be an example of an untamed black swan. As I mentioned in the section on solar flares, we have observed superflares on other stars that are 10,000 bigger than the biggest flare we have seen from the Sun, so far.

According to the log-t model, the probability of a flare that size is a few per billion, which means we should expect one every 30,000 years, on average. But that estimate extrapolates beyond the data by four powers of 10; we have little confidence that the model is valid in that range. Maybe the processes that generate superflares are different from the processes that generate normal flares, and they are actually more likely than the model predicts. Or maybe our Sun is different from stars that produce superflares, and it will never produce a superflare, ever. A model based on 20 years of data can't distinguish among these possibilities; they are unknown unknowns.

The difference between black and gray swans is the state of our knowledge and our best models. If a superflare happened tomorrow, it would be a black swan. But as we learn more about the Sun and other stars, we might create better models that predict large solar flares with credible accuracy. In the future, a superflare might be a gray swan. However, predicting rare events will always be difficult, and there may always be black swans.

IN A LONG-TAILED WORLD

In fact, there are two reasons we will always have a hard time dealing with rare, large events. First, they are rare, and people are bad at comprehending small probabilities. Second, they are large, and people are bad at comprehending very large magnitudes. To demonstrate the second point, let's imagine a world where the distribution of human height is not Gaussian or even lognormal; instead, it follows a long-tailed distribution like the various disasters in this chapter. Specifically, suppose it follows a log-t distribution with the same tail thickness as the distribution of disaster costs, but with the same mean as the distribution of human height.

If you woke up in this Long-Tailed World, you might not notice the difference immediately. The first few people you meet would probably be of unsurprising height. Out of 100 people, the tallest might be 215 cm, which is quite tall, but you would not suspect, yet, that you have been transported to a strange new world. If you see 1000 people, the tallest of them might be 277 cm, and you might start to suspect something strange. Out of 10,000 people, the tallest would be more than four meters tall; at that point, you would know you were not in Kansas.

But we're just getting started. Out of a million people in Long-Tailed World, the tallest would be 59 meters, about the height of a eucalyptus tree. In a country the size of the United States, the tallest person would be 160,000 kilometers tall; standing on the surface of the earth, their head would be more than a third of the way to the moon. And in a world population of seven billion, the tallest would be 14 quintillion kilometers tall, which is about 1500 light years, almost three times the distance from our Sun to Betelgeuse.

This example shows how long-tailed distributions violate our intuition and strain our imagination. Most of the things we observe directly in the world follow Gaussian and lognormal distributions. The things that follow long-tailed distributions are hard to perceive, which makes them hard to conceive. However, by collecting data and using appropriate tools to understand it, we can make better predictions and—I hope—prepare accordingly. Otherwise, rare events will continue to take us by surprise and catch us unprepared.

In this chapter I made it a point to identify the sources of the data I used; I am grateful to the following agencies and organizations for making this data freely available:

- The Space Weather Prediction Center and the National Centers for Environmental Information, funded by the National Oceanic and Atmospheric Administration (NOAA)
- The Southern California Earthquake Data Center, funded by the US Geological Survey (USGS)
- The Lunar Crater Database, provided by the USGS using data collected by NASA
- The JPL Small-Body Database, provided by the Jet Propulsion Laboratory and funded by NASA
- The MeasuringWorth Foundation, a nonprofit organization with the mission to make available "the highest quality and most reliable historical data on important economic aggregates"

Collecting data like this requires satellites, space probes, seismometers, and other instruments; and processing it requires time, expertise, and other resources. These activities are expensive, but if we use the data to predict, prepare for, and mitigate the costs of disaster, it will pay for itself many times over.

SOURCES AND RELATED READING

- The list of disasters is from Wikipedia [70].
- John D. Cook wrote a blog post about Student's *t* distribution [21].
- An informative video on solar flares and coronal mass ejections is available from Kurzgesagt [23].
- The data on solar flares is available from the Space Weather Prediction Center Data Service [120]; at SpaceWeatherLive, you can see a video of the largest flare in the dataset, on September 7, 2005 [9], and photos of the resulting aurora at SpaceWeather.com [112].
- A coronal mass ejection was the cause of the Carrington Event, the most intense geomagnetic storm in recorded history [15].
- The earthquake data is available from the Southern California Earthquake Data Center [115].
- The asteroid that exploded near Greenland was reported in the *New York Times* [6].
- The meteor explosion over Siberia is known as the Tunguska Event [126].
- The moon crater data is described in the *Journal of Geophysical Research: Planets* [102] and available from the Lunar Crater Database [101].
- The asteroid data is available from the JPL Small-Body Database [56].
- I got the relationship between the sizes of craters and the sizes of asteroids from Melosh, *Planetary Surface Processes* [78].
- Benoit Mandelbrot's heretical explanation of long-tailed distributions is from *The Fractal Geometry of Nature* [73].
- The stock market data is available from MeasuringWorth [132].
- Nassim Nicholas Taleb's discussion of black and gray swans is in *The Black Swan* [121].

CHAPTER 9

FAIRNESS AND FALLACY

In the criminal justice system, we use algorithms to guide decisions about who should be released on bail or kept in jail and who should be kept in prison or released on parole. Of course, we want those algorithms to be fair. For example, suppose we use an algorithm to predict whether a candidate for parole is likely to commit another crime, if released. If the algorithm is fair,

- its predictions should mean the same thing for different groups of people. For example, if the algorithm predicts that a group of women and a group of men are equally likely to re-offend, we expect the numbers of women and men who actually re-offend to be the same, on average.
- Also, since the algorithm will make mistakes—some people who re-offend will be assigned low probabilities, and some people who do not re-offend will be assigned high probabilities—it should make these mistakes at the same rate in different groups of people.

It is hard to argue with either of these requirements. Suppose the algorithm assigns a probability of 30% to 100 women and 100 men; if 20 of the women commit another crime and 40 of the men do, it seems like the algorithm is unfair. Or suppose Black candidates are more likely to be wrongly assigned a high probability and White candidates are more likely to be wrongly assigned a low probability. That seems unfair, too.

But here's the problem: it is not possible for an algorithm—or a human—to satisfy both requirements. Unless two groups commit crimes at precisely the same rate, any classification that is equally predictive for both groups will necessarily make different kinds of errors between the groups. And if we calibrate it to make the same kind of errors, the meaning of the predictions will be different. To understand why, we have to understand the *base rate fallacy*.

In this chapter, I'll demonstrate the base rate fallacy using three examples:

- Suppose you take a medical test and the result is positive, which indicates that you have a particular disease. If the test is 99% accurate, you might think there is a 99% chance that you have the disease. But the actual probability could be much lower.
- Suppose a driver is arrested because a testing device found that their blood alcohol was above the legal limit. If the device is 99% accurate, you might think there's a 99% chance they are guilty. In fact, the probability depends strongly on the reason the driver was stopped.
- Suppose you hear that 70% of people who die from a disease had been vaccinated against it. You might think that the vaccine was not effective. In fact, such a vaccine might prevent 80% of deaths or more and save a large number of lives.

These examples might be surprising if you have not seen them before, but once you understand what's going on, I think they will make sense. At the end of the chapter, we'll return to the problem of algorithms and criminal justice.

MEDICAL TESTS

During the COVID-19 pandemic, we all got a crash course in the science and statistics of infectious disease. One of the things we learned about is the accuracy of medical testing and possibility of errors, both false positives and false negatives. As an example, suppose a friend tells you they have tested positive for COVID, and they want to know whether they are really infected or whether the result might be a false positive. What information would you need to answer their question?

One thing you clearly need to know is the accuracy of the test, but that turns out to be a little tricky. In the context of medical testing, we have to consider two kinds of accuracy: sensitivity and specificity.

- *Sensitivity* is the ability of the test to detect the presence of an infection, usually expressed as a probability. For example, if the sensitivity of the test is 87%, that means 87 out of 100 people who are actually infected will get a positive test result, on average. The other 13 will get a *false negative*.
- *Specificity* is the ability of the test to indicate the absence of an infection. For example, if the specificity is 98%, that means 98 out of 100 people who are not infected will get a negative test result, on average. The other two will get a *false positive*.

I did not make those numbers up. They are the reported sensitivity of a particular rapid antigen test, the kind used for at-home testing, in December 2021. By the numbers, it sounds like the test is accurate, so if your friend tested positive, you might think they are likely to be infected.

But that's not necessarily true. It turns out that there is another piece of information we need to consider: the *base rate*, which is the probability that your friend was infected, based on everything we know about them *except* the outcome of the test. For example, if they live someplace where the infection rate is high, we know they have been in a room with someone who was infected, and they currently have symptoms, the base rate might be quite high. If they have been in strict isolation for 14 days and have no symptoms, it would be quite low.

To see why it matters, let's consider a case where the base rate is relatively low, like 1%. And let's imagine a group of 1000 people who all take the test. In a group this size, we expect 10 people to be infected, because 10 out of 1000 is 1%. Of the 10 who are actually infected, we expect 9 to get a positive test result, because the sensitivity of the test is 87%. Of the other 990, we expect 970 to get a negative test result, because the specificity is 98%. But that means we expect 20 people to get a false positive.

Before we go on, let's put the numbers we have so far in a table.

	# of people	Prob positive	# positive
Infected	10	0.87	9
Not infected	990	0.02	20

The first column is the number of people in each group: infected and not infected. The second column is the probability of a positive test for each group. For someone who is actually infected, the probability of a positive test is 0.87, because sensitivity is 87%. For someone who is not infected, the probability of a *negative* test is 0.98, because specificity is 98%, so the probability of a positive test is 0.02. The third column is the product of the first two columns, which is the number of positive tests we expect in each group, on average. Out of 1000 test results, 9 are true positives and 20 are false positives, for a total of 29.

Now we are ready to answer your friend's question: Given a positive test result, what is the probability that they are actually infected? In this example, the answer is 9 out of 29, or 31%. Here's the table again, with a fourth column showing the probability of actual infection and the complementary probability that the test result is a false positive.

	# of people	Prob positive	# positive	% true positive
Infected	10	0.87	9	31.0
Not infected	990	0.02	20	69.0

Although the sensitivity and specificity of the test are high, after a positive result, the probability that your friend is infected is only 31%. The reason it's so low is that the base rate in this example is only 1%.

MORE PREVALENCE

To see why it matters, let's change the scenario. Suppose your friend has mild, flu-like symptoms; in that case it seems more likely that they are infected, compared to someone with no symptoms. Let's say it is 10 times more likely, so the probability that your friend is infected is 10% before we get the test results. In that case, out of

1000 people with the same symptoms, we would expect 100 to be infected. If we modify the first column of the table accordingly, here are the results.

	# of people	Prob positive	# positive	% true positive
Infected	100	0.87	87	82.9
Not infected	900	0.02	18	17.1

Now the probability is about 83% that your friend is actually infected and about 17% that the result is a false positive. This example demonstrates two things:

1. the base rate makes a big difference, and
2. even with an accurate test and a 10% base rate, the probability of a false positive is still surprisingly high.

If the test is more sensitive, that helps, but maybe not as much as you expect. For example, another brand of rapid antigen tests claims 95% sensitivity, substantially better than the first brand, which was 87%. With this test, assuming the same specificity, 98%, and the same base rate, 10%, here's what we get:

	# of people	Prob positive	# positive	% True positive
Infected	100	0.95	95	84.1
Not infected	900	0.02	18	15.9

Increasing the sensitivity from 87% to 95% has only a small effect: the probability that the test result is a false positive goes from 17% to 16%.

MORE SPECIFICITY

Increasing specificity has a bigger effect. For example, lab tests that use PCR (polymerase chain reaction) are highly specific, about as close to 100% as can be. However, in practice it is always possible that a specimen is contaminated, a device malfunctions, or a result is reported incorrectly.

For example, in a retirement community near my house in Massachusetts, 18 employees and one resident tested positive for COVID in August 2020. But all 19 turned out to be false positives, produced by a lab in Boston that was suspended by the Department of Public Health after they reported at least 383 false positive results.

It's hard to say how often something like that goes wrong, but if it happens one time in 1000, the specificity of the test would be 99.9%. Let's see what effect that has on the results.

	# of people	Prob positive	# positive	% True positive
Infected	100	0.950	95.0	99.1
Not infected	900	0.001	0.9	0.9

With 95% sensitivity, 99.9% specificity, and 10% base rate, the probability is about 99% that your friend is actually infected, given a positive PCR test result.

However, the base rate still matters. Suppose you tell me that your friend lives in New Zealand, where (at least at the time I am writing) the rate of COVID infection is very low. In that case the base rate for someone with mild, flu-like symptoms might be one in 1000. Here's the table with 95% sensitivity, 99.9% specificity, and base rate one in 1000.

	# of people	Prob positive	# positive	% true positive
Infected	1	0.950	0.950	48.7
Not infected	999	0.001	0.999	51.3

In this example, the numbers in the third column aren't integers, but that's okay. The calculation works the same way. Out of 1000 tests, we expect 0.95 true positives, on average, and 0.999 false positives. So the probability is about 49% that a positive test is correct. That's lower than most people think, including most doctors.

BAD MEDICINE

A 2014 paper in the *Journal of the American Medical Association* reports the result of a sneaky experiment. The researchers asked a "convenience sample" of doctors (probably their friends and colleagues) the following question: "If a test to detect a disease whose prevalence is 1/1000 has a false positive rate of 5%, what is the chance that a person found to have a positive result actually has the disease, assuming you know nothing about the person's symptoms or signs?" What they call "prevalence" is what I've been calling "base rate." And what they call the "false positive rate" is the complement of specificity, so a false positive rate of 5% corresponds to a sensitivity of 95%.

Before I tell you the results of the experiment, let's work out the answer to the question. We are not given the sensitivity of the test, so I'll make the optimistic assumption that it is 99%. The following table shows the results.

	# of people	Prob positive	# positive	% true positive
Infected	1	0.99	0.99	1.9
Not infected	999	0.05	49.95	98.1

The correct answer is about 2%.

Now, here are the results of the experiment: "Approximately three-quarters of respondents answered the question incorrectly. In our study, 14 of 61 respondents (23%) gave a correct response. [. . .] The most common answer was 95%, given by 27 of 61 respondents." If the correct answer is 2% and the most common response is 95%, that is an alarming level of misunderstanding.

To be fair, the wording of the question might have been confusing. Informally, "false positive rate" could mean either

- the fraction of uninfected people who get a positive test result, or
- the fraction of positive test results that are false.

The first is the technical definition of "false positive rate"; the second is called the "false discovery rate." But even statisticians have trouble keeping these terms straight, and doctors are experts at medicine, not statistics.

However, even if the respondents misunderstood the question, their confusion could have real consequences for patients. In the case of COVID testing, a false positive test result might lead to an unnecessary period of isolation, which would be disruptive, expensive, and possibly harmful. A state investigation of the lab that produced hundreds of false positive results concluded that their failures "put patients at immediate risk of harm."

Other medical tests involve similar risks. For example, in the case of cancer screening, a false positive might lead to additional tests, unnecessary biopsy or surgery, and substantial costs, not to mention emotional difficulty for the patient and their family. Doctors and patients need to know about the base rate fallacy. As we'll see in the next section, lawyers, judges, and jurors do, too.

DRIVING UNDER THE INFLUENCE

The challenges of the base rate fallacy have become more salient as some states have cracked down on "drugged driving." In September 2017 the American Civil Liberties Union (ACLU) filed suit against Cobb County, Georgia on behalf of four drivers who were arrested for driving under the influence of cannabis. All four were evaluated by Officer Tracy Carroll, who had been trained as a "Drug Recognition Expert" (DRE) as part of a program developed by the Los Angeles Police Department in the 1970s.

At the time of their arrest, all four insisted that they had not smoked or ingested any cannabis products, and when their blood was tested, all four results were negative; that is, the blood tests found no evidence of recent cannabis use. In each case, prosecutors dismissed the charges related to impaired driving. Nevertheless, the arrests were disruptive and costly, and the plaintiffs were left with a permanent and public arrest record.

At issue in the case is the assertion by the ACLU that "much of the DRE protocol has never been rigorously and independently validated." So I investigated that claim. What I found was a collection of studies that are, across the board, deeply flawed. Every one of them features at least one methodological error so blatant it would be embarrassing at a middle school science fair.

As an example, the lab study most often cited to show that the DRE protocol is valid was conducted at Johns Hopkins University School of Medicine in 1985. It concludes, "Overall, in 98.7% of instances of judged intoxication the subject had received some active drug." In other words, in the cases where one of the Drug Recognition Experts believed that a subject was under the influence, they were right 98.7% of the time.

That sounds impressive, but there are several problems with this study. The biggest is that the subjects were all "normal, healthy" male volunteers between 18 and 35 years old, who were screened and "trained on the psychomotor tasks and subjective effect questionnaires used in the study." By design, the study excluded women, anyone older than 35, and anyone in poor health. Then the screening excluded anyone who had any difficulty passing a sobriety test while they were sober—for example, anyone with shaky hands, poor coordination, or poor balance. But those are exactly the people most likely to be falsely accused. How can you estimate the number of false positives if you exclude from the study everyone likely to yield a false positive? You can't.

Another frequently cited study reports that "when DREs claimed drugs other than alcohol were present, they [the drugs] were almost always detected in the blood (94% of the time)." Again, that sounds impressive until you look at the methodology. Subjects in this study had already been arrested because they were suspected of driving while impaired, most often because they had failed a field sobriety test. Then, while they were in custody, they were evaluated by a DRE, that is, a different officer trained in the drug evaluation procedure. If the DRE thought that the suspect was under the influence of a drug, the suspect was asked to consent to a blood test; otherwise they were released.

Of 219 suspects, 18 were released after a DRE performed a "cursory examination" and concluded that there was no evidence of drug impairment. The remaining 201 suspects were asked for a blood sample. Of those, 22 refused and 6 provided a urine sample only. Of the 173 blood samples, 162 were found to contain a drug other than alcohol. That's about 94%, which is the statistic they reported.

But the base rate in this study is extraordinarily high, because it includes only cases that were suspected by the arresting officer and then confirmed by the DRE. With a few generous assumptions, I estimate that the base rate is 86%; in reality, it was probably higher. Because the suspects who were released were not tested, there is no way to estimate the sensitivity of the test, but let's assume it's 99%, so if a suspect is under the influence of a drug, there is a 99% chance a DRE would detect it. In reality, it is probably lower. With these generous assumptions, we can use the following table to estimate the sensitivity of the DRE protocol.

	Suspects	Prob positive	Cases	%
Impaired	86	0.99	85.14	93.8
Not impaired	14	0.40	5.60	6.2

With an 86% base rate, out of 100 suspects, we expect 86 impaired and 14 unimpaired. With 99% sensitivity, we expect the DRE to detect about 85 true positives. And with 60% specificity, we expect the DRE to wrongly accuse 5.6 suspects. Out of 91 positive tests, 85 would be correct; that's about 94%, as reported in the study. But this accuracy is only possible because the base rate in the study is so high. Remember that most of the subjects had been arrested because they had failed a field sobriety test. Then they were tested by a DRE, who was effectively offering a second opinion.

But that's not what happened when Officer Tracy Carroll arrested Katelyn Ebner, Princess Mbamara, Ayokunle Oriyomi, and Brittany Penwell. In each of those cases, the driver was stopped for driving erratically, which is evidence of possible impairment. But when Officer Carroll began his evaluation, that was the only evidence of impairment.

So the relevant base rate is not 86%, as in the study; it is the fraction of erratic drivers who are under the influence of drugs. And there are many other reasons for erratic driving, including distraction, sleepiness, and the influence of alcohol. It's hard to say which explanation is most common. I'm sure it depends on time and location. But as an example, let's suppose it is 50%; the following table shows the results with this base rate.

	Suspects	Prob positive	Cases	%
Impaired	50	0.99	49.5	71.2
Not impaired	50	0.40	20.0	28.8

With 50% base rate, 99% sensitivity, and 60% specificity, the predictive value of the test is only 71%; under these assumptions, almost 30% of the accused would be innocent. In fact, the base rate, sensitivity, and specificity are probably lower, which means that the value of the test is even worse.

The suit filed by the ACLU was not successful. The court decided that the arrests were valid because the results of the field sobriety tests constituted "probable cause" for an arrest. As a result, the court did not consider the evidence for, or against, the validity of the DRE protocol. The ACLU has appealed the decision.

VACCINE EFFECTIVENESS

Now that we understand the base rate fallacy, we're ready to untangle a particularly confusing example of COVID disinformation. In October 2021, a journalist appeared on a well-known podcast with a surprising claim. He said, "In the [United Kingdom] 70-plus percent of the people who die now from COVID are fully vaccinated." The incredulous host asked, "70 percent?" and the journalist repeated, "Seven in ten of the people [who died]—I want to keep saying because nobody believes it but the numbers are there in the government documents—the vast majority of people in Britain who died in September were fully vaccinated."

Then, to his credit, he showed a table from a report published by Public Health England in September 2021. From the table, he read off the number of deaths in each age group: "1270 out of 1500 in the over 80 category, [. . .] 607 of 800 of the 70 year-olds. [. . .] They were almost all fully vaccinated. Most people who die of this now are fully vaccinated in the UK. Those are the numbers." It's true—those are the numbers. But the implication that the vaccine is useless, or actually harmful, is wrong. In fact, we can use these numbers, along with additional information from the same table, to compute the effectiveness of the vaccine and estimate the number of lives it saved.

Let's start with the oldest age group, people who were 80 or more years old. In this group, there were 1521 deaths attributed to COVID during the four-week period from August 23 to September 19, 2021. Of the people who died, 1272 had been fully vaccinated. The others were either unvaccinated or partially vaccinated; for simplicity I'll consider them all not fully vaccinated. So, in this age group, 84% of the people who died had been fully vaccinated. On the face of it, that sounds like the vaccine was not effective.

However, the same table also reports death rates among the vaccinated and unvaccinated, that is, the number of deaths as a fraction of the population in each age group. During the same four-week period, the death rates due to COVID were 1560 per million people among the unvaccinated and 495 per million among the vaccinated. So, the death rate was substantially lower among the vaccinated.

The following table shows these death rates in the second column and the number of deaths in the third column. Given these numbers, we can work forward to compute the fourth column, which shows again that 84% of the people who died had been vaccinated. We can also work backward to compute the first column, which shows that there were about 2.57 million people in this age group who had been vaccinated and only 0.16 million who had not. So, more than 94% of this age group had been vaccinated.

	Population	Death rate	Deaths	%
Vaccinated	2.57	495	1272	83.6
Not vaccinated	0.16	1560	249	16.4

From this table, we can also compute the effectiveness of the vaccine, which is the fraction of deaths the vaccine prevented. The difference in the death rate from 1560 per million to 495 per million is a decrease of 68 percentage points. By definition, this decrease is the "effectiveness" of the vaccine in this age group.

Finally, we can estimate the number of lives saved by answering a counterfactual question: if the death rate among the vaccinated had been the same as the death rate among the unvaccinated, how many deaths would there have been? The answer is that there would have been 4009 deaths. In reality, there were 1272, so we can estimate that the vaccine saved about 2737 lives in this age group, in just four weeks. In the United Kingdom right now, there are a lot of people visiting parents and grandparents at their homes, rather than a cemetery, because of the COVID vaccine.

Of course, this analysis is based on some assumptions, most notably that the vaccinated and unvaccinated were similar except for their vaccination status. That might not be true: people with high risk or poor general health might have been more likely to seek out the vaccine. If so, our estimate would be too low, and the vaccine might have saved more lives. If not, and people in poor health were *less* likely to be vaccinated, our estimate would be too high. I'll leave it to you to judge which is more likely.

We can repeat this analysis with the other age groups. The following table shows the number of deaths in each age group and the percentage of the people who died who were vaccinated. I've omitted the "under 18" age group because there were only six deaths in this group, four among the unvaccinated and two with unknown status. With such small numbers, we can't make useful estimates for the death rate or effectiveness of the vaccine.

Age	Deaths vax	Deaths total	Vax percentage of deaths
18 to 29	5	17	29
30 to 39	10	48	21
40 to 49	30	104	29
50 to 59	102	250	41
60 to 69	258	411	63
70 to 79	607	801	76
80+	1272	1521	84

Adding up the columns, there were a total of 2284 deaths, 3152 of them among the vaccinated. So 72% of the people who died had been vaccinated, as the unnamed journalist reported. Among people over 80, it was even higher, as we've already seen. However, in the younger age groups, the percentage of deaths among the vaccinated is substantially lower, which is a hint that this number might reflect something about the groups, not about the vaccine.

To compute something about the vaccine, we can use death rates rather than number of deaths. The following table shows death rates per million people, reported by Public Health England for each age group, and the implied effectiveness of the vaccine, which is the percent reduction in death rate.

Age	Death rate vax	Death rate unvax	Effectiveness
18 to 29	1	3	67
30 to 39	2	12	83
40 to 49	5	38	87
50 to 59	14	124	89
60 to 69	45	231	81
70 to 79	131	664	80
80+	495	1560	68

The effectiveness of the vaccine is more than 80% in most age groups. In the youngest group it is 67%, but that might be inaccurate because the number of deaths is low, and the estimated death rates are not precise. In the oldest group, it is 68%, which suggests that the vaccine is less effective for older people, possibly because their immune systems are weaker. However, a treatment that reduces the probability of dying by 68% is still very good. Effectiveness is nearly the same in most age groups because it reflects primarily something about the vaccines and only secondarily something about the groups.

Now, given the number of deaths and death rates, we can infer the number and percentage of people in each age group who were vaccinated.

Age	Vax total (millions)	Unvax total (millions)	Vax percentage
18 to 29	5.0	4.0	56
30 to 39	5.0	3.2	61
40 to 49	6.0	1.9	75
50 to 59	7.3	1.2	86
60 to 69	5.7	0.7	90
70 to 79	4.6	0.3	94
80+	2.6	0.2	94

By August 2021, nearly everyone in the England over 60 years old had been vaccinated. In the younger groups, the percentages were lower, but even in the youngest group it was more than half. With this, it becomes clear why most deaths were among the vaccinated:

- most deaths were in the oldest age groups, and
- in those age groups, almost everyone was vaccinated.

Taking this logic to the extreme, if everyone is vaccinated, we expect all deaths to be among the vaccinated.

In the vocabulary of this chapter, the percentage of deaths among the vaccinated depends on the effectiveness of the vaccine *and* the base rate of vaccination in the population. If the base rate is low, as in the younger groups, the percentage of deaths among the vaccinated is low. If the base rate is high, as in the older groups, the percentage of deaths is high. Because this percentage depends so strongly on the properties of the group, it doesn't tell us much about the properties of the vaccine.

Finally, we can estimate the number of lives saved in each age group. First we compute the hypothetical number of the vaccinated who would have died if their death rate had been the same as among the unvaccinated, then we subtract off the actual number of deaths.

Age	Hypothetical deaths	Actual deaths	Lives saved
18 to 29	15	5	10
30 to 39	60	10	50
40 to 49	228	30	198
50 to 59	903	102	801
60 to 69	1324	258	1066
70 to 79	3077	607	2470
80+	4009	1272	2737

In total, the COVID vaccine saved more than 7000 lives in a four-week period, in a relevant population of about 48 million. If you created a vaccine that saved 7000 lives in less than a month, in just one country, you would feel pretty good about yourself. And if you used misleading statistics to persuade a large, international audience that they should not get that vaccine, you should feel very bad.

PREDICTING CRIME

If we understand the base rate fallacy, we can correctly interpret medical and impaired driving tests, and we can avoid being misled by headlines about COVID vaccines. We can also shed light on an ongoing debate about the use of data and algorithms in the criminal justice system.

In 2016 a team of journalists at *ProPublica* published a now-famous article about COMPAS, which is a statistical tool used in some states to inform decisions about which defendants should be released on bail before trial, how long convicted defendants should be imprisoned, and whether prisoners should be released on probation. COMPAS uses information about defendants to generate a "risk score" that is supposed to quantify the probability that the defendant will commit another crime if released.

The authors of the *ProPublica* article used public data to assess the accuracy of COMPAS risk scores. They explain, "We obtained the risk scores assigned to more than 7000 people arrested in Broward County, Florida, in 2013 and 2014 and checked to see how many were charged with new crimes over the next two years, the same benchmark used by the creators of the algorithm." They published the data they obtained, so we can use it to replicate their analysis and do our own.

If we think of COMPAS as a diagnostic test, a high risk score is like a positive test result and a low risk score is like a negative result. Under those definitions, we can use the data to compute the sensitivity and specificity of the test. As it turns out, they are not very good.

- *Sensitivity*: Of the people who were charged with another crime during the period of observation, only 63% were given high risk scores.
- *Specificity*: Of the people who were *not* charged with another crime, only 68% were given low risk scores.

Now suppose you are a judge considering a bail request from a defendant who has been assigned a high risk score. Among other things, you would like to know the probability that they will commit a crime if released. Let's see if we can figure that out. As you might guess by now, we need another piece of information: the base rate. In the sample from Broward County, it is 45%; that is, 45% of the defendants released from jail were charged with a crime within two years.

The following table shows the results with this base rate, sensitivity, and specificity.

	# of people	P (high risk)	# high risk	%
Charged again	450	0.63	283	61.8
Not charged	550	0.32	175	38.2

Out of 1000 people in this dataset, 450 will be charged with a crime, on average; the other 550 will not. Based on the sensitivity and specificity of the test, we expect 283 of the offenders to be assigned a high risk score, along with 175 of the non-offenders. So, of all people with high risk scores, about 62% will be charged with another crime. This result is called the "positive predictive value," or PPV, because it quantifies the accuracy of a positive test result. In this case, 62% of the positive tests turn out to be correct.

We can do the same analysis with low risk scores.

	# of people	P (low risk)	# low risk	Percent
Charged again	450	0.37	166	30.7
Not charged	550	0.68	374	69.3

Out of 450 offenders, we expect 166 to get an incorrect low score. Out of 550 non-offenders, we expect 374 to get a correct low score. So, of all people with low risk scores, 69% were not charged with another crime. This result is called the "negative predictive value" of the test, or NPV, because it indicates what fraction of negative tests are correct.

On one hand, these results show that risk scores provide useful information. If someone gets a high risk score, the probability is 62% that they will be charged with a crime. If they get a low risk score, it is only 31%. So, people with high risk scores are about twice as likely to re-offend. On the other hand, these results are not as accurate as we would like when we make decisions that affect people's lives so seriously. And they might not be fair.

COMPARING GROUPS

The authors of the *ProPublica* article considered whether COMPAS has the same accuracy for different groups. With respect to racial groups, they find:

> . . . In forecasting who would re-offend, the algorithm made mistakes with Black and White defendants at roughly the same rate but in very different ways.
>
> - The formula was particularly likely to falsely flag Black defendants as future criminals, wrongly labeling them this way at almost twice the rate as White defendants.
> - White defendants were mislabeled as low risk more often than Black defendants.

This discrepancy suggests that the use of COMPAS in the criminal justice system is racially biased.

I will use the data they obtained to replicate their analysis, and we will see that their numbers are correct. But interpreting these results turns out to be complicated; I think it will be clearer if we start by considering first sex and then race. In the data from Broward County, 81% of defendants are male and 19% are female. The sensitivity and specificity of the risk scores is almost the same in both groups:

- Sensitivity is 63% for male defendants and 61% for female defendants.
- Specificity is close to 68% for both groups.

But the base rate is different: about 47% of male defendants were charged with another crime, compared to 36% of female defendants.

In a group of 1000 male defendants, the following table shows the number we expect to get a high risk score and the fraction of them that will re-offend.

	# of people	P (high risk)	# high risk	%
Charged again	470	0.63	296	63.7
Not charged	530	0.32	169	36.3

Of the high-risk male defendants, about 64% were charged with another crime. Here is the corresponding table for 1000 female defendants.

	# of people	P (high risk)	# high risk	%
Charged again	360	0.61	219	51.8
Not charged	640	0.32	204	48.2

Of the high-risk female defendants, only 52% were charged with another crime. And that's what we should expect: If the test has the same sensitivity and specificity, but the groups have different base rates, the test will have different predictive values in the two groups.

Now let's consider racial groups. As the *ProPublica* article reports, the sensitivity and specificity of COMPAS are substantially different for White and Black defendants:

- Sensitivity for White defendants is 52%; for Black defendants it is 72%.
- Specificity for White defendants is 77%; for Black defendants it is 55%.

The complement of sensitivity is the "false negative rate," or FNR, which in this context is the fraction of offenders who were wrongly classified as low risk. The false negative rate for White defendants is 48% (the complement of 52%); for Black defendants it is 28%. And the complement of specificity is the "false positive rate," or FPR, which is the fraction of non-offenders who were wrongly classified as high risk. The false positive rate for White defendants is 23% (the complement of 77%); for Black defendants, it is 45%.

In other words, Black non-offenders were almost twice as likely to bear the cost of an incorrect high score. And Black offenders were substantially less likely to get the benefit of an incorrect low score. That seems patently unfair. As US Attorney General Eric Holder wrote in 2014 (as quoted in the *ProPublica* article), "Although these measures were crafted with the best of intentions, I am concerned that they inadvertently undermine our efforts to ensure individualized and equal justice [and] they may exacerbate unwarranted and unjust disparities that are already far too common in our criminal justice system and in our society."

But that's not the end of the story.

FAIRNESS IS HARD TO DEFINE

A few months after the *ProPublica* article, the *Washington Post* published a response with the expositive title "A computer program used for bail and sentencing decisions was labeled biased against blacks. It's actually not that clear." It acknowledges that the results of the *ProPublica* article are correct: the false positive rate for Black defendants is higher, and the false negative rate is lower. But it points out that the PPV and NPV are the nearly the same for both groups:

- *Positive predictive value*: Of people with high risk scores, 59% of White defendants and 63% of Black defendants were charged with another crime.
- *Negative predictive value*: Of people with low risk scores, 71% of White defendants and 65% of Black defendants were *not* charged again.

So in this sense the test is fair: a high risk score in either group means the same thing; that is, it corresponds to roughly the same probability of recidivism. And a low risk score corresponds to roughly the same probability of non-recidivism.

Strangely, COMPAS achieves one kind of fairness based on sex and another kind of fairness based on race.

- For male and female defendants, the error rates (false positive and false negative) are roughly the same, but the predictive values are different.
- For Black and White defendants, the error rates are substantially different, but the predictive values (PPV and NPV) are about the same.

The COMPAS algorithm is a trade secret, so there is no way to know why it is designed this way, or even whether the discrepancy is deliberate. But the discrepancy is not inevitable. COMPAS could be calibrated to have equal error rates in all four groups, or equal predictive values. However, it cannot have the same error rates *and* the same predictive values. We have already seen why: If the error rates are the same and the base rates are different, we get different predictive values. And, going the other way, if the predictive values are the same and the base rates are different, we get different error rates.

At this point it is tempting to conclude that algorithms are inherently unfair, so maybe we should rely on humans instead of algorithms. But this option is not as appealing as it might seem. First, it doesn't actually solve the problem: just like an algorithm, a human judge cannot achieve equal error rates for different groups and equal predictive values at the same time. The math is inescapable by man or machine.

Second, if the task is to use data to generate predictions, humans are almost always worse than algorithms. To see why, let's consider the reasons a person and an algorithm might disagree:

1. A human might consider additional information that's not available to the algorithm. For example, a judge might assess whether a defendant seems contrite, based on their behavior in court.
2. A human might consider the same information as an algorithm, but give different weight to different factors. For example, a judge might give more weight to age than the algorithm does and less weight to prior arrests.
3. A human might be influenced by factors that don't affect algorithms, like political beliefs, personal biases, and mood.

Taking these in turn:

1. If a human judge uses more information than the algorithm, the additional information may or may not be valid. If not, it provides no advantage. If so, it could be included in the algorithm. For example, if judges record their belief about whether a defendant is contrite or not, we could check whether their assessment is actually predictive, and if so, we could add it to the algorithm.
2. If a human judge gives more weight to some factors, relative to the algorithm, and less weight to others, the results are unlikely to be better.

After all, figuring out which factors are predictive and how much weight to give each one, is exactly what algorithms are designed to do, and they are generally better at it than humans.

3. Finally, there is ample evidence that judges differ from each other in consistent ways, and that they differ from themselves over time. The outcome of a case should not depend on whether it is assigned to a harsh or a lenient judge, or whether it is heard before or after lunch. And it certainly should not depend on prejudices the judge may have based on race, sex, and other group membership.

I don't mean to say that algorithms are guaranteed to be free of this kind of prejudice. If they are based on previous outcomes, and if those outcomes are subject to bias, algorithms can replicate and perpetuate that bias. For example, the dataset used by *ProPublica* to validate COMPAS indicates whether each defendant was *charged* with another crime during the period of observation. But what we really want to know is whether the defendant *committed* another crime, and that is not the same thing.

Not everyone who commits a crime gets charged—not even close. The probability of getting charged for a particular crime depends on the type of crime and location; the presence of witnesses and their willingness to work with police; the decisions of police about where to patrol, what crimes to investigate, and who to arrest; and decisions of prosecutors about who to charge. It is likely that every one of these factors depends on the race and sex of the defendant.

This kind of data bias is a problem for algorithms like COMPAS. But it is also a problem for humans: exposed to biased data, we tend to make biased judgments. The difference is that humans can handle less data, and we are less good at extracting reliable information from it. Trained with the same data, an algorithm will be about as biased as the average judge, less biased than the worst judge, and less noisy than any judge.

Also, algorithms are easier to correct than humans. If we discover that an algorithm is biased, and we can figure out how, we can often unbias it. If we could do that with humans, the world would be a better place. For all of these reasons, I think algorithms like COMPAS have a place in the criminal justice system. But that brings us back to the question of calibration.

FAIRNESS IS HARD TO ACHIEVE

Even if you think we should not use predictive algorithms in the criminal justice system, the reality is that we do. So at least for now we have some difficult questions to answer:

1. Should we calibrate algorithms so predictive values are the same in all groups and accept different error rates (as we see with Black and White defendants)?
2. Or should we calibrate them so error rates are the same in all groups and accept different predictive values (as we see with male and female defendants)?
3. Or should we compromise between the extremes and accept different error rates *and* different predictive values?

If we choose either of the first two options, we run into two problems: the number of groups is large, and every defendant belongs to several of them. Consider a defendant who is a 50-year old African American woman. What is the false positive rate for her group? As we've already seen, FPR for Black defendants is 45%. But for Black women it's 40%; for women older than 45, it's 15%; and for Black women older than 45, it's 24%. We have the same problem with the false negative rate. For example, FNR for White defendants is 48%, but for White women, it is 43%; for women younger than 25, it's 18%; and for White women younger than 25, it's just 4%!

Predictive values (PPV and NPV) don't differ as much between groups, but if you search for the extremes, you can find substantial differences. Among the subgroups I looked at (excluding very small groups),

- COMPAS has the highest positive predictive value for Black men younger than 25, 70%. It has the lowest PPV for Hispanic defendants older than 45, 29%.

- It has the highest negative predictive value for White women younger than 25, 95%, and the lowest NPV for men under 25 whose racial category is "other," 49%.

With six racial categories, three age groups, and two sexes, there are 81 subgroups. It is not possible to calibrate any algorithm to achieve the same error rates or the same predictive values in all of these groups.

So, suppose we use an algorithm that allows error rates and predictive values to vary between groups. How should we design it, and how should we evaluate it? Let me suggest a few principles to start with:

1. If one of the goals of incarceration is to reduce crime, it is better to keep in prison someone who will commit another crime, if released, than someone who will not. Of course we don't know with certainty who will re-offend, but we can make probabilistic predictions.

2. The public interest is better served if our predictions are accurate, otherwise we will keep more people in prison than necessary, or suffer more crime than necessary, or both.

3. However, we should be willing to sacrifice some accuracy in the interest of justice. For example, suppose we find that, comparing male and female defendants who are alike in every other way, women are more likely to re-offend. In that case, including sex in the algorithm might improve its accuracy. Nevertheless, we might decide to exclude this information on the grounds that using it would violate the principle of equality before the law.

4. The criminal justice system should be fair, and it should be perceived to be fair. However, we have seen that there are conflicting definitions of fairness, and it is mathematically impossible to satisfy all of them.

Even if we agree that these principles should guide our decisions, they provide a framework for a discussion rather than a resolution. For example, reasonable people could disagree about what factors should be included in the algorithm. I suggested that sex should be excluded even if it improves the accuracy of the predictions. For the same reason, we might choose to exclude race.

But what about age? If two defendants are similar except that one is 25 years old and the other is 50, the younger person is substantially more likely to re-offend. So an algorithm that includes age will be more accurate than one that does not. And on the face of it, releasing someone from prison because they are old does not seem obviously unjust. But a person does not choose their age any more than they choose their race or sex. So I'm not sure what principle justifies the decision to include age while excluding race and sex.

The point of this example is that these decisions are hard because they depend on values that are not universal. Fortunately, we have tools for making decisions when people disagree, including public debate and representative democracy. But the key words in that sentence are "public" and "representative." The algorithms we use in the criminal justice system should be a topic of public discussion, not a trade secret. And the debate should include everyone involved, including perpetrators and victims of crime.

ALL ABOUT THE BASE RATE

Sometimes the base rate fallacy is funny. There's a very old joke that goes like this: "I read that 21% of car crashes are caused by drunk drivers. Do you know what that means? It means that 79% are caused by sober drivers. Those sober drivers aren't safe—get them off the road!" And sometimes the base rate fallacy is obvious, as in the *xkcd* comic that says, "Remember, right-handed people commit 90% of all base rate errors."

But often it is more subtle. When someone says a medical test is accurate, they usually mean that it is sensitive and specific: that is, likely to be positive if the condition it detects is present, and likely to be negative if the condition is absent. And those are good properties for a test to have. But they are not enough to tell us what we really want to know, which is whether a particular result is correct. For that, we need the base rate, and it often depends on the circumstances of the test.

For example, if you go to a doctor because you have symptoms of a particular disease, and they test for the disease, that's a diagnostic test. If the test is sensitive and specific, and the result is positive, it's likely that you have the disease. But if you go to the doctor for a regular checkup, you have no symptoms, and they test for a rare disease, that's a screening test. In that case, if the result is positive, the probability that you have the disease might be small, even if the test is highly specific. It's important for you to know this, because there's a good chance your doctor does not.

SOURCES AND RELATED READING

- The sensitivity and specificity of COVID antigen tests was reported in the "Wirecutter" feature of the *New York Times* [64].
- The false positive COVID tests from the Orig3n lab were reported in the *Boston Globe* [80].
- The experiment that asked doctors to interpret the results of a medical test was described in *JAMA Internal Medicine* [74].
- The story of the drivers in Georgia who were wrongly arrested for driving under the influence of cannabis was reported by *11 Alive*, WXIA-TV, in Atlanta, Georgia [58].
- The complaint filed on their behalf and updates on the appeal are available from the ACLU website [36].
- The 1985 lab test and the DRE protocol are reported in a technical report from the National Highway Traffic Safety Administration [14]. The 1986 field test is reported in another technical report [20].
- The podcast where the unnamed journalist committed the base rate fallacy is *The Joe Rogan Experience* [39].
- The data he reported is from Public Health England [24].
- The *ProPublica* article about COMPAS and the response in the *Washington Post* are available online [7] [22].
- My discussion of machine bias is based on a case study that is part of the *Elements of Data Science* curriculum [33]. The Python code I used to analyze the data is available in Jupyter notebooks [34].
- For a discussion of the vagaries of human judges, I recommend chapter 1 of *Noise: A Flaw in Human Judgment* [57].

- The company that produces COMPAS was called Northpointe in 2016. After the *ProPublica* article was published, the name of the company was changed to Equivant.

- The *xkcd* comic about the base rate fallacy is number 2476 [82]. The hover text is "Sure, you can talk about per-capita adjustment, but if you want to solve the problem, it's obvious that this is the group you need to focus on."

- A January 2022 feature in the "Upshot" blog at the *New York Times* discusses the high false positive rates of screening tests for rare genetic disorders [60].

- A 2022 article in the *New York Times* discusses potential harms of increased screening for rare cancers [61].

CHAPTER 10

PENGUINS, PESSIMISTS, AND PARADOXES

In 2021, the journalist I mentioned in the previous chapter wrote a series of misleading articles about the COVID pandemic. I won't name him, but on April 1, appropriately, the *Atlantic* magazine identified him as "The Pandemic's Wrongest Man." In November, this journalist posted an online newsletter with the title "Vaccinated English adults under 60 are dying at twice the rate of unvaccinated people the same age." It includes a graph showing that the overall death rate among young, vaccinated people increased between April and September, 2021, and was actually higher among the vaccinated than among the unvaccinated.

As you might expect, this newsletter got a lot of attention. Among skeptics, it seemed like proof that vaccines were not just ineffective, but harmful. And among people who were counting on vaccines to end the pandemic, it seemed like a devastating setback. Many people fact-checked the article, and at first the results held up to scrutiny: the graph accurately represented data published by the UK Office for National Statistics. Specifically, it showed death rates, from any cause, for people in England between 10 and 59 years old, from March to September 2021. And between April and September, these rates were higher among the fully vaccinated, compared to the unvaccinated, by a factor of almost two. So the journalist's description of the results was correct.

However, his conclusion that vaccines were causing increased mortality was completely wrong. In fact, the data he reported shows that vaccines were safe and effective in this population and that they demonstrably saved many lives. How, you might ask, can increased mortality be evidence of saved lives? The answer is Simpson's paradox. If you have not seen it before, Simpson's paradox can seem impossible, but as I will try to show you, it is not just possible but ordinary—and once you've seen enough examples, it is not even surprising. We'll start with an easy case and work our way up.

OLD OPTIMISTS, YOUNG PESSIMISTS

Would you say that most of the time people try to be helpful, or that they are mostly just looking out for themselves? Almost every year since 1972, the General Social Survey (GSS) has posed that question to a representative sample of adult residents of the United States. The following figure shows how the responses have changed over time. The circles show the percentage of people in each survey who said that people try to be helpful.

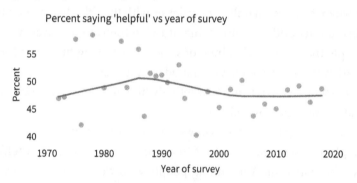

The percentages vary from year to year, in part because the GSS includes different people in each survey. Looking at the extremes: in 1978, about 58% said "helpful"; in 1996, only 40% did. The solid line in the figure shows a LOWESS curve, which is a statistical way to smooth out short-term variation and show the long-term trend. Putting aside the extreme highs and lows, it looks like the fraction of optimists has decreased since 1990, from about 51% to 48%.

These results don't just depend on when you ask; they also depend on who you ask. In particular, they depend strongly on when the respondents were born, as shown in the following figure.

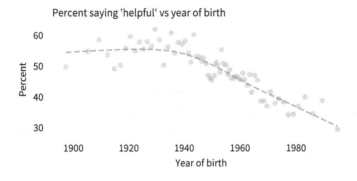

Percent saying 'helpful' vs year of birth

Notice that the *x*-axis here is the respondent's year of birth, not the year they participated in the survey. The oldest person to participate in the GSS was born in 1883. Because the GSS only surveys adults, the youngest, as of 2021, was born in 2003. Again, the markers show variability in the survey results; the dashed line shows the long-term trend.

Starting with people born in the 1940s, Americans' faith in humanity has been declining consistently, from about 55% for people born before 1930 to about 30% for people born after 1990. That is a substantial decrease! To get an idea of what's happening, let's zoom in on people born in the 1940s. The following figure shows the fraction of people in this cohort who said people would try to be helpful, plotted over time.

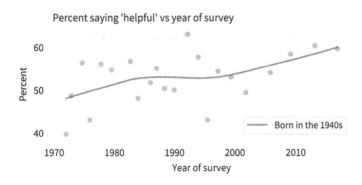

Percent saying 'helpful' vs year of survey

Again, different people were surveyed each year, so there is a lot of variability. Nevertheless, there is a clear increasing trend. When this cohort was interviewed in the 1970s, about 46% of them gave a positive response; when they were interviewed in the 2010s, about 61% did. So that's surprising.

So far, we've seen that

- people have become more pessimistic over time,
- and successive generations have become more pessimistic,
- but if we follow one cohort over time, they become more optimistic as they age.

Now, maybe there's something unusual about people born in the 1940s. They were children during the aftermath of World War II and they were young adults during the rise of the Soviet Union and the Cold War. These experiences might have made them more distrustful when they were interviewed in the 1970s. Then, from the 1980s to the present, they enjoyed decades of relative peace and increasing prosperity. And when they were interviewed in the 2010s, most were retired and many were wealthy. So maybe that put them in a more positive frame of mind.

However, there is a problem with this line of conjecture: it wasn't just one cohort that trended toward the positive over time. It was all of them. The following figure shows the responses to this question grouped by decade of birth and plotted over time. For clarity, I've dropped the circles showing the annual results and plotted only the smoothed lines. The solid lines show the trends within each cohort: people generally become more optimistic as they age. The dashed line shows the overall trend: on average, people are becoming less optimistic over time.

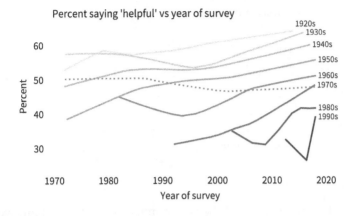

This is an example of Simpson's paradox, named for Edward H. Simpson, a codebreaker who worked at Bletchley Park during World War II. For people who have not seen it before, the effect often seems impossible: if all of the groups are increasing, and we combine them into a single group, how can the aggregate decrease? In this example, the explanation is *generational replacement*; as the oldest, more optimistic cohort dies off, it is replaced by the youngest, more pessimistic cohort.

The differences between these generations are substantial. In the most recent data, only 39% of respondents born in the 1990s said people would try to be helpful, compared to 51% of those born in the 1960s and 64% of those born in the 1920s. And the composition of the population has changed a lot over 50 years. The following figure shows the distribution of birth years for the respondents at the beginning of the survey in 1973, near the middle in 1990, and most recently in 2018.

Distribution of birth year

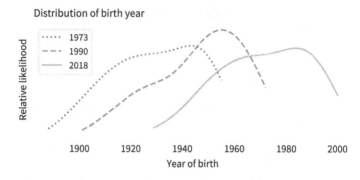

In the earliest years of the survey, most of the respondents were born between 1890 and 1950. In the 1990 survey, most were born between 1901 and 1972. In the 2018 survey, most were born between 1929 and 2000.

With this explanation, I hope Simpson's paradox no longer seems impossible. But if it is still surprising, let's look at another example.

REAL WAGES

In April 2013 Floyd Norris, chief financial correspondent of the *New York Times*, wrote an article with the headline "Median Pay in U.S. Is Stagnant, but Low-Paid Workers Lose." Using data from the US Bureau of Labor Statistics, he computed the following changes in median real wages between 2000 and 2013, grouped by level of education:

- no high school degree: −7.9%
- high school degree, no college: −4.7%
- some college: −7.6%
- bachelor's degree or higher: −1.2%

In all groups, real wages were lower in 2013 than in 2000, which is what the article was about. But Norris also reported the overall change in the median real wage: during the same period, it *increased* by 0.9%. Several readers wrote to say he had made a mistake. If wages in all groups were going down, how could overall wages go up?

To explain what's going on, I will replicate the results from Norris's article, using data from the Current Population Survey (CPS) conducted by the US Census Bureau and the Bureau of Labor Statistics. This dataset includes wages and education levels for almost 1.9 million participants between 1996 and 2021. Wages are adjusted for inflation and reported in constant 1999 dollars. The following figure shows average real wages for each level of education over time.

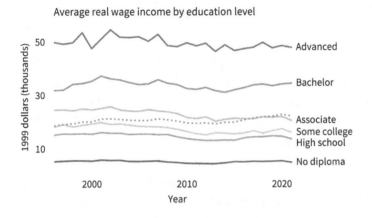

These results are consistent with Norris's: real wages in every group decreased over this period. The decrease was steepest among people with an associate degree, by about $190 per year over 26 years. The decrease was smallest among people with no high school diploma, about $30 per year. However, if we combine the groups, real wages increased during the same period by about $80 per year, as shown by the dotted line in the figure. Again, this seems like a contradiction.

In this example, the explanation is that levels of education changed substantially during this period. The following figure shows the fraction of the population at each educational level, plotted over time.

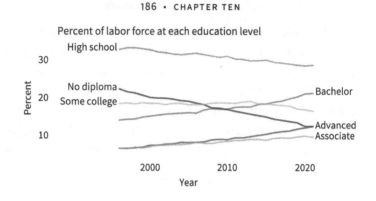

Compared to 1996, people in 2021 have more education. The groups with associate degrees, bachelor's degrees, or advanced degrees have grown by 3–7 percentage points. Meanwhile, the groups with no diploma, high school, or some college have shrunk by 2–10 percentage points.

So the overall increase in real wages is driven by changes in education levels, not by higher wages at any level. For example, a person with a bachelor's degree in 2021 makes less money, on average, than a person with a bachelor's degree in 1996. But a random person in 2021 is more likely to have a bachelor's degree (or more) than a person in 1996, so they make more money, on average.

In both examples so far, the explanation for Simpson's paradox is that the population is a mixture of groups and the proportions of the mixture change over time. In the GSS examples, the population is a mixture of generations, and the older generations are replaced by younger generations. In the CPS example, the population is a mixture of education levels, and the fraction of people at each level changes over time.

But Simpson's paradox does not require the variable on the x-axis to be time. More generally, it can happen with any variables on the x- and y-axes and any set of groups. To demonstrate, let's look at an example that doesn't depend on time.

PENGUINS

In 2014 a team of biologists published measurements they collected from three species of penguins living near the Palmer Research Station in Antarctica. For 344 penguins, the dataset includes body mass, flipper length, and the length and depth of their beaks: more specifically, the part of the beak called the culmen. This dataset has become popular with machine-learning experts because it is a fun way to demonstrate all kinds of statistical methods, especially classification algorithms.

Allison Horst, an environmental scientist who helped make this dataset freely available, used it to demonstrate Simpson's paradox. The following figure shows the example she discovered.

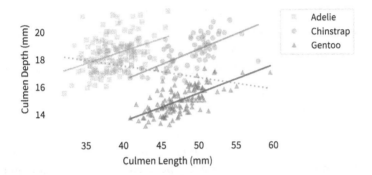

Each marker represents the culmen length and depth of an individual penguin. The different marker shapes represent different species: Adélie, Chinstrap, and Gentoo. The solid lines are the lines of best fit within each species; in all three, there is a positive correlation between length and depth. The dotted line shows the line of best fit among all of the penguins; between species, there is a negative correlation between length and depth.

In this example, the correlation within species has some meaning: most likely, length and depth are correlated because both are correlated with body size. Bigger penguins have bigger beaks. And it is significant that the different species are clustered around different points. As Darwin famously observed among Galápagos finches, the dimensions of a bird's beak are generally adapted to their ecological niche. If different penguin species have different diets, their beaks likely reflect that specialization.

However, when we combine samples from different species and

find that the correlation of these variables is negative, that's a statistical artifact. It depends on the species we choose and how many penguins of each species we measure. So it doesn't really mean anything.

Penguins might be interesting, but dogs take Simpson's paradox to a new level. In 2007, a team of biologists at Leiden University in the Netherlands and the Natural History Museum in Berne, Switzerland described the unusual case of one Simpson's paradox nested inside another. They observed:

- When we compare species, there is a positive correlation between weight and lifespan: In general, bigger animals live longer.
- However, when we compare dog breeds, there is a negative correlation: Bigger breeds have shorter lives, on average.
- However, if we compare dogs within a breed, there is a positive correlation again: A bigger dog is expected to live longer than a smaller dog of the same breed.

In summary, elephants live longer than mice, Chihuahuas live longer than Great Danes, and big beagles live longer than small beagles. The trend is different at the level of species, breed, and individual because the relationship at each level is driven by different biological processes. These results are puzzling only if you think the trend within groups is necessarily the same as the trend between groups. Once you realize that's not true, the paradox is resolved.

SIMPSON'S PRESCRIPTION

Some of the most bewildering examples of Simpson's paradox come up in medicine, where a treatment might prove to be effective for men, effective for women, and ineffective or harmful for people. To understand examples like that, let's start with an analogous example from the General Social Survey. Since 1973, the GSS has asked respondents, "Is there any area right around here—that is, within a mile—where you would be afraid to walk alone at night?" Out of more than 37,000 people who have answered this question, about 38% said, "Yes." Since 1974, they have also asked, "Should divorce

in this country be easier or more difficult to obtain than it is now?"
About 49% of the respondents said it should be more difficult.

Now, if you had to guess, would you expect these responses to be
correlated; that is, do you think someone who says they are afraid
to walk alone at night might be more likely to say divorce should be
more difficult? As it turns out, they are slightly correlated:

- Of people who said they are afraid, about 50% said divorce should be
  more difficult.
- Of people who said they are not afraid, about 49% said divorce should be
  more difficult.

However, there are several reasons we should not take this correlation too seriously:

- The difference between the groups is too small to make any real difference, even if it were valid.
- It combines results from 1974 to 2018, so it ignores the changes in both
  responses over that time.
- It is actually an artifact of Simpson's paradox.

To explain the third point, let's look at responses from men and
women separately. The following figure shows this breakdown.

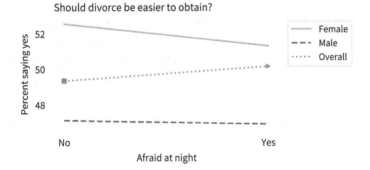

Among women, people who are afraid are less likely to say divorce
should be difficult. And among men, people who are afraid are *also*

less likely to say divorce should be difficult (very slightly). In both groups, the correlation is negative, but when we put them together, the correlation is positive. People who are afraid are more likely to say divorce should be difficult.

As with the other examples of Simpson's paradox, this might seem to be impossible. But as with the other examples, it is not. The key is to realize that there is a correlation between the sex of the respondent and their answers to both questions. In particular, of the people who said they are not afraid, 41% are female; of the people who said they *are* afraid, 74% are female. You can actually see these proportions in the figure:

- On the left, the square marker is 41% of the way between the male and female proportions.
- On the right, the triangle marker is 74% of the way between the male and female proportions.

In both cases, the overall proportion is a mixture of the male and female proportions. But it is a *different* mixture on the left and right. That's why Simpson's paradox is possible.

With this example under our belts, we are ready to take on one of the best-known examples of Simpson's paradox. It comes from a paper published in 1986 by researchers at St. Paul's Hospital in London who compared two treatments for kidney stones: (A) open surgery and (B) percutaneous nephrolithotomy. For our purposes, it doesn't matter what these treatments are; I'll just call them A and B.

Among 350 patients who received Treatment A, 78% had a positive outcome; among another 350 patients who received Treatment B, 83% had a positive outcome. So it seems like Treatment B is better.

However, when they split the patients into groups, they found that the results were reversed:

- Among patients with relatively small kidney stones, Treatment A was better, with a success rate of 93% compared to 86%.
- Among patients with larger kidney stones, Treatment A was better again, with a success rate of 73% compared to 69%.

The following figure shows these results in graphical form.

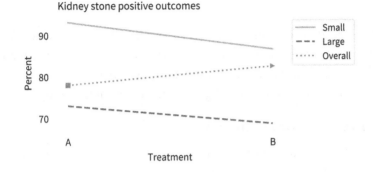

This figure resembles the previous one, because the explanation is the same in both cases: the overall percentages, indicated by the square and triangle markers, are mixtures of the percentages from the two groups. But they are different mixtures on the left and right.

For medical reasons, patients with larger kidney stones were more likely to receive Treatment A; patients with smaller kidney stones were more likely to receive Treatment B. As a result, among people who received Treatment A, 75% had large kidney stones, which is why the square marker on the left is closer to the Large group. And among people who received Treatment B, 77% had small kidney stones, which is why the triangle marker is closer to the Small group.

I hope this explanation makes sense, but even if it does, you might wonder how to choose a treatment. Suppose you are a patient and your doctor explains the following things:

- For someone with a small kidney stone, Treatment A is better.
- For someone with a large kidney stone, Treatment A is better.
- But overall, Treatment B is better.

Which treatment should you choose? Other considerations being equal, you should choose A because it has a higher chance of success for any patient, regardless of stone size. The apparent success of B is a statistical artifact, the result of two causal relationships:

- Smaller stones are more likely to get Treatment B.
- Smaller stones are more likely to lead to a good outcome.

Looking at the overall results, Treatment B only looks good because it is used for easier cases; Treatment A only looks bad because it is used for the hard cases. If you are choosing a treatment, you should look at the results within the groups, not between the groups.

DO VACCINES WORK? HINT: YES

Now that we have mastered Simpson's paradox, let's get back to the example from the beginning of the chapter, the newsletter that went viral with the claim that "vaccinated English adults under 60 are dying at twice the rate of unvaccinated people the same age."

The newsletter includes a graph based on data from the UK Office for National Statistics. I downloaded the same data and replicated the graph, which looks like this:

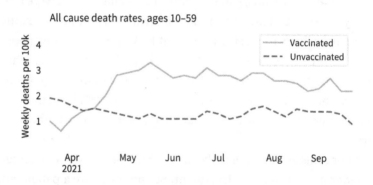

The lines show weekly death rates from all causes, per 100,000 people age 10–59, from March to September 2021. The solid line represents people who were fully vaccinated; the dashed line represents people who were unvaccinated.

The journalist I won't name concludes, "Vaccinated people under 60 are twice as likely to die as unvaccinated people. And overall deaths in Britain are running well above normal. I don't know how to explain this other than vaccine-caused mortality." Well, I do. There are two things wrong with his interpretation of the data:

- First, by choosing one age group and time interval, he has cherry-picked data that support his conclusion and ignored data that don't.
- Second, because the data combine a wide range of ages, from 10 to 59, he has been fooled by Simpson's paradox.

Let's debunk one thing at a time. First, here's what the graph looks like if we include the entire dataset, which starts in January:

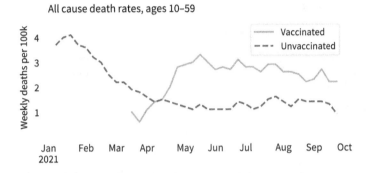

All cause death rates, ages 10–59

Overall death rates were "running well above normal" in January and February, when almost no one in the 10–59 age range had been vaccinated. Those deaths cannot have been caused by the vaccine; in fact, they were caused by a surge in the Alpha variant of COVID-19.

The unnamed journalist leaves out the time range that contradicts him; he also leaves out the age ranges that contradict him. In all of the older age ranges, death rates were consistently lower among people who had been vaccinated. The following figures show death rates for people in their 60s and 70s and for people over 80.

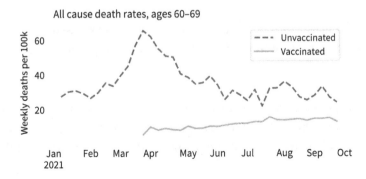

All cause death rates, ages 60–69

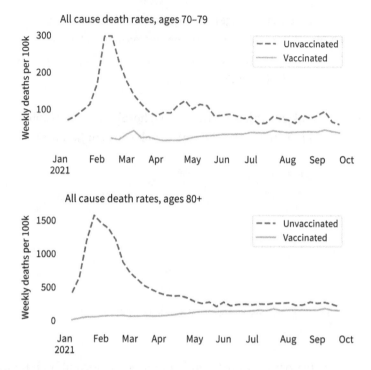

Notice that the *y*-axes are on different scales; death rates are much higher in the older age groups. In all of these groups, death rates are substantially lower for people who are vaccinated, which is what we expect from a vaccine that has been shown in large clinical trials to be safe and effective.

So what explains the apparent reversal among people between 10 and 59 years old? It is a statistical artifact caused by Simpson's paradox. Specifically, it is caused by two correlations:

- Within this age group, older people were more likely to be vaccinated.
- Older people were more likely to die of any cause, just because they are older.

Both of these correlations are strong. For example, at the beginning of August, about 88% of people at the high end of this age range had been vaccinated; *none* of the people at the low end had. And the death rate for people at the high end was 54 times higher than for people at the low end.

To show that these correlations are strong enough to explain the observed difference in death rates, I will follow an analysis by epidemiologist Jeffrey Morris, who refuted the unreliable journalist's claims within days of their publication. He divides the excessively wide 10–59 year age group into 10 groups, each five years wide.

To estimate normal, pre-pandemic death rates in these groups, he uses 2019 data from the UK Office of National Statistics. To estimate the vaccination rate in each group, he uses data from the coronavirus dashboard provided by the UK Health Security Agency. Finally, to estimate the fraction of people in each age group, he uses data from a web site called PopulationPyramid, which organizes and publishes data from the United Nations Department of Economic and Social Affairs. Not to get lost in the details, these are all reliable data sources.

Combining these datasets, Morris is able to compute the distribution of ages in the vaccinated and unvaccinated groups at the beginning of August 2021 (choosing a point near the middle of the interval in the original graph). The following figure shows the results.

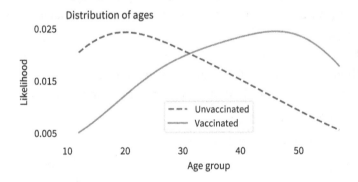

At this point during the rollout of the vaccines, people who were vaccinated were more likely to be at the high end of the age range, and people who were unvaccinated were more likely to be at the low end. Among the vaccinated, the average age was about 40; among the unvaccinated, it was 27.

Now, for the sake of this example, let's imagine that there are no deaths due to COVID and no deaths due to the vaccine. Based only on the distribution of ages and the death rates from 2019, we can compute the expected death rates for the vaccinated and unvaccinated groups. The ratio of these two rates is about 2.4. So, prior to the pandemic, we would expect people with the age distribution of the vaccinated to die at a rate 2.4 times higher than people with the age distribution of the unvaccinated, just because they are older.

In reality, the pandemic increased the death rate in both groups, so the actual ratio was smaller, about 1.8. Thus, the results the journalist presented are not evidence that vaccines are harmful; as Morris concludes, they are "not unexpected, and can be fully explained by the Simpson's paradox artifact."

DEBUNK REDUX

So far we have used only data that was publicly available in November 2021, but if we take advantage of more recent data, we can get a clearer picture of what was happening then and what has happened since. In more recent reports, data from the UK Office of National Statistics are broken into smaller age groups. Instead of one group from ages 10 to 59, we have three groups: from 18 to 39, 40 to 49, and 50 to 59.

Let's start with the oldest of these subgroups. Among people in their 50s, we see that the all-cause death rate was substantially higher among unvaccinated people over the entire interval from March 2021 to April 2022.

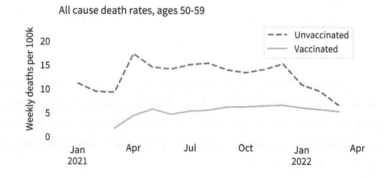

If we select people in their 40s, we see the same thing: death rates were higher for unvaccinated people over the entire interval.

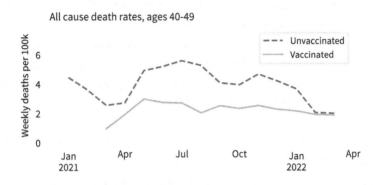

All cause death rates, ages 40-49

However, it is notable that death rates in both groups have declined since April or May 2021 and have nearly converged. Possible explanations include improved treatment for coronavirus disease and the decline of more lethal variants of the COVID virus. In the UK, December 2021 is when the Delta variant was largely replaced by the Omicron variant, which is less deadly.

Finally, let's see what happened in the youngest group, people aged 18 to 39.

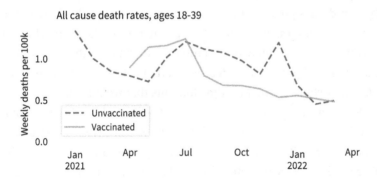

All cause death rates, ages 18-39

In this age group, like the others, death rates are generally higher among unvaccinated people. There are a few months where the pattern is reversed, but we should not take it too seriously, for the following reasons:

- Death rates in this age group are very low, less than one per 100,000 per week, vaccinated or not.
- The apparent differences between the groups are even smaller and might be the result of random variation.
- This age group is still broad enough that the results are influenced by Simpson's paradox. The vaccinated people in this group were older than the unvaccinated, and the older people in this group were about 10 times more likely to die, just because of their age.

In summary, the excess mortality among vaccinated people in 2021 is entirely explained by Simpson's paradox. If we break the dataset into appropriate age groups, we see that death rates were higher among the unvaccinated in all age groups over the entire time from the rollout of the vaccine to the present. And we can conclude that the vaccine saved many lives.

OPEN DATA, OPEN DISCUSSION

Reflecting on this episode, we can identify several elements that went wrong. A journalist downloaded data he did not have the expertise to interpret correctly. He wrote about it in a newsletter that was published with no editorial review. Its spread was accelerated by a credulous podcaster who ignored numerous easily discoverable articles debunking its claims. And, given the circulation it achieved, it contributed to a body of vaccine misinformation that almost certainly caused preventable deaths. All that is true; nevertheless, it is also important to consider the elements that went right.

First, the UK government collected reliable data about the pandemic and made it widely available. Other countries followed their example, albeit some more successfully than others. Groups like Our World in Data collected similar data from around the world and made it easier to use, which accelerated the ability of researchers to understand the pandemic, make predictions, and guide public policy decisions. This data saved lives.

Second, after the misleading newsletter was posted on November 20, dozens of articles debunked its claims within days. When I ran a web search for the newsletter, I immediately found rebuttals published on November 21, 22, and 23.

Finally, even on social media sites that are infamous for spreading disinformation, public discussion of the article was not all bad. Simpson's paradox is not easy to understand, but some of the explanations were very good. I think it's a hopeful sign to see people grappling with statistical ideas in a public forum.

The marketplace of ideas will never be perfect, but this episode, which shows its weaknesses, also shows its strengths. We are better off when our decisions are guided by evidence and reason, but that is only possible if we have access to data and open spaces to discuss difficult ideas.

SOURCES AND RELATED READING

- The General Social Survey (GSS) is a project of the University of Chicago's independent research organization NORC, with primary financing from the National Science Foundation. Data is available from the GSS website [44].
- Floyd Norris wrote about changes in real wages in the *New York Times* [86]. The following week, he wrote a blog article to respond to readers who thought he had made a mistake [85].
- The penguin dataset was reported in "Ecological Sexual Dimorphism and Environmental Variability within a Community of Antarctic Penguins" [49]. Allison Horst pointed out the penguin paradox in the online documentation of the dataset [53].
- The nested Simpson's paradox among dog breeds is described in "Do Large Dogs Die Young?" [43].
- The paper that evaluates treatments for kidney stones was published in 1986 in the *British Medical Journal* [16].
- Derek Thompson wrote about the misleading journalist in the *Atlantic* [123]. Jeffrey Morris debunks the misleading newsletter in a series of blog articles [81].
- Information about the prevalence of different COVID variants is from the UK Health Security Agency [109].
- The COVID data I used is from the UK Office for National Statistics [29].

CHAPTER 11

CHANGING HEARTS AND MINDS

In 1950, the physicist Max Planck made a bleak assessment of progress in science. He wrote, "A new scientific truth does not triumph by convincing its opponents and making them see the light, but rather because its opponents eventually die, and a new generation grows up that is familiar with it." Scientists often quote a pithier version attributed to economist Paul A. Samuelson: "Science progresses, one funeral at a time." According to this view, science progresses by generational replacement alone, not changed minds.

I am not sure that Planck and Samuelson are right about science, but I don't have the data to check. However, thanks to the General Social Survey (GSS), we have the data to assess a different kind of progress, the expansion of the "moral circle." The idea of the moral circle was introduced by historian William Lecky in *A History of European Morals from Augustus to Charlemagne*, published in 1867. He wrote, "At one time the benevolent affections embrace merely the family, soon the circle expanding includes first a class, then a nation, then a coalition of nations, then all humanity, and finally, its influence is felt in the dealings of man with the animal world."

In this chapter we'll use data from the GSS to explore the moral circle, focusing on questions related to race, sex, and sexual orientation. We'll find more examples of Simpson's paradox, which we saw in the previous chapter. For example, older people are more likely to hold racist views, but that doesn't mean people become more racist as they get older. To interpret this result and others like it, I'll introduce a tool called age-period-cohort analysis and a concept called the Overton window.

Let's start with race.

OLD RACISTS?

Stereotypes suggest that older people are more racist than young people. To see whether that's true, I'll use responses to three questions in the General Social Survey related to race and public policy:

1. Do you think there should be laws against marriages between (Negroes/Blacks/African-Americans) and whites?
2. If your party nominated a (Negro/Black/African-American) for President, would you vote for him if he were qualified for the job?
3. Suppose there is a community-wide vote on the general housing issue. There are two possible laws to vote on. Which law would you vote for?
   - One law says that a homeowner can decide for himself whom to sell his house to, even if he prefers not to sell to [people of a particular race].
   - The second law says that a homeowner cannot refuse to sell to someone because of their race or color.

I chose these questions because they were added to the survey in the early 1970s, and they have been asked almost every year since. The words in parentheses indicate that the phrasing of these questions has changed over time to use contemporary terms for racial categories.

The following figure shows the responses to these questions as a function of the respondents' ages. To make it easy to compare answers to different questions, the *y*-axis shows the percentage who chose what I characterize as the racist responses: that interracial marriage should be illegal; that the respondent would not vote for a Black presidential candidate; and that it should be legal to refuse to sell a house to someone based on their race. The results vary from year to year, so I've plotted a smooth curve to fit the data.

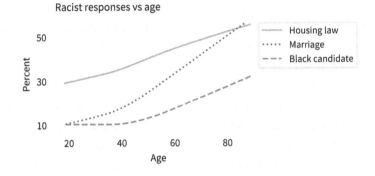

For all three questions, old people are substantially more likely to choose the racist response, so there is some truth to the stereotype. But that raises a question: do people *become* more racist as they age, or do they persist in the beliefs they were raised with? We can answer that with another view of the data. The following figure shows responses to the first question, about interracial marriage, grouped by decade of birth and plotted by age.

The dotted line shows the trend for all respondents; as we saw in the previous figure, older respondents are more likely to favor a law against interracial marriage. But that doesn't mean that people are more likely to adopt this view as they age. In fact, it's the opposite: in almost every birth cohort, people age out of racism.

So this is another example of Simpson's paradox: within the groups, the trend is downward as a function of age, but overall, the trend is upward. The reason is that, because of the design of the GSS, we observe different cohorts at different ages. At the left side of the figure, the overall average is low because the youngest people surveyed are mostly from the most recent generations; at the right side, the overall average is high because the older people are mostly from the earliest generations.

You might notice that we have only one observation for people born in the 1980s and no data for people born in the 1990s. That's because this question was dropped from the GSS after 2002. At that point, the percentage of people in favor of the law, and willing to say so, had dropped below 10%. Among people born in the 1980s, it was barely 1%.

At that point, the GSS had several reasons to retire the question. First, as a matter of public policy, the matter was resolved in 1967 by the US Supreme Court decision in *Loving v. Virginia*; by the 1990s it was no longer part of mainstream political discussion. Second, because the responses were so one-sided, there was little to learn by asking. And finally, estimating small proportions from survey data is unreliable due to a phenomenon known as "lizard people."

The term comes from a notorious poll conducted in 2013 by Public Policy Polling, which included questions about a variety of conspiracy theories. One asked, "Do you believe that shape-shifting reptilian people control our world by taking on human form and gaining political power to manipulate our societies, or not?" Of 1247 registered voters who responded, 4% said yes. If that is an accurate estimate of the prevalence of this belief, it implies that there are more than 12 million people in the United States who believe in lizard people.

But it is probably not an accurate estimate, because of a problem well known to survey designers. In any group of respondents, there will be some percentage who misunderstand a question, accidentally choose a response they did not intend, or maliciously choose a response they do not believe. And in this example, there were probably a few open-minded people who had never heard of lizard people in positions of power, but once the survey raised the possibility, they were willing to entertain it.

Errors like this are tolerable when the actual prevalence is high. In that case, the number of errors is small compared to the number of legitimate positive responses. But when the actual prevalence is low, there might be more false positive responses than true ones. If 4% of respondents endorsed the lizard people theory in error, the actual prevalence might be zero. The point of this diversion is that it is hard to measure small percentages with survey data, which is one reason the GSS stops asking about rare beliefs.

The responses to the other two questions follow the same pattern:

- Asked whether they would vote for a qualified Black presidential candidate nominated by their own party, older people were more likely to say no. But within every birth cohort, people were more likely to say yes as they got older.
- Asked whether they would support an open housing law, older people were more likely to say no. But within every birth cohort, people were more likely to say yes as they got older.

So, even if you observe that older people are more likely to hold racist beliefs, that doesn't mean people become more racist with age. In fact, the opposite is true: in every generation, going back to 1900, people grew less racist over time.

YOUNG FEMINISTS

Similarly, older people are more likely to hold sexist beliefs, but that doesn't mean people become more sexist as they age. The GSS includes three questions related to sexism:

1. Please tell me whether you strongly agree, agree, disagree, or strongly disagree [. . .]: It is much better for everyone involved if the man is the achiever outside the home, and the woman takes care of the home and family.
2. Tell me if you agree or disagree with this statement: Most men are better suited emotionally for politics than are most women.
3. If your party nominated a woman for President, would you vote for her if she were qualified for the job?

The following figure shows the results as a function of the respondents' ages. Again, the y-axis shows the percentage of respondents who chose what I characterize as a sexist response: that it is much better for everyone if women stay home, that men are more suited emotionally for politics, and that the respondent would not vote for a female presidential candidate.

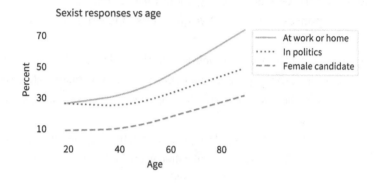

For all three questions, older people are more likely to choose the sexist response. The difference is most dramatic for the first question, related to women working outside the home. The difference is smaller for the other questions, although that is in part because the third question, related to voting for a female presidential candidate, was retired after 2010 as the prevalence dropped into lizard people territory. The following figure shows responses to the second question grouped by decade of birth and plotted by age.

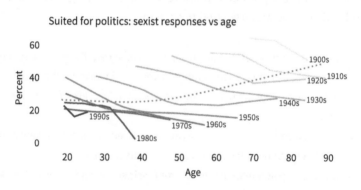

Overall, older people are more likely to be sexist, but within almost every cohort, people become less sexist as they get older. The results from the other two questions show the same pattern. As in the previous examples, the reason is that the younger respondents are more likely to belong to recent birth cohorts, which are less sexist, and the older respondents are more likely to belong to earlier birth cohorts, which are more sexist.

This example demonstrates the difference between an "age effect" and a "cohort effect"; in general, there are three ways populations change over time:

- An "age effect" is something that affects most people at a particular age or life stage. Age effects can be biological, like the loss of deciduous teeth, or social, like the loss of youthful innocence.
- A "period effect" is something that affects most people at a particular point in time. Period effects include notable events, like the September 11 attacks, or intervals like the Cold War.
- A "cohort effect" is something that affects people born at a particular time, usually due to the environment they grew up in. When people make generalizations about the characteristics of baby boomers and millennials, for example, they are appealing to cohort effects (often without much evidence).

It can be hard to distinguish between these effects. In particular, when we see a cohort effect, it is easy to mistake it for an age effect. When we see that older people hold particular beliefs, we might assume (or fear) that young people will adopt those beliefs as they age. But age effects like that are rare.

Most people develop social beliefs based on the environment they are raised in, so that's primarily a cohort effect. If those beliefs change over their lifetimes, it is most often because of something happening in the world, which is a period effect. But it is unusual to adopt or change a belief when you reach a particular age. Other than registering to vote when you are 18, most political acts don't depend on the number of candles on the cake.

The previous figure shows one way to distinguish cohort and age effects: grouping people by cohort and plotting their responses as a function of age. Similarly, to distinguish cohort and period effects, we can group people by birth cohort, again, and plot their responses over time. For example, the following figure shows responses to the first of the three questions, about whether everyone would be better off if women stayed home, plotted over time.

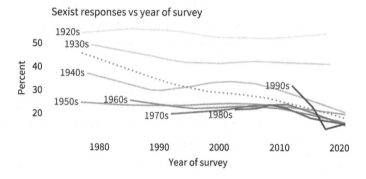

In this figure, there is evidence of a cohort effect: comparing people born in the 1920s through the 1950s, each cohort is less likely to choose a sexist response than the previous one. There is also some evidence for a weak period effect. Among the cohorts born in the 1950s through the 1990s, the percentage of sexist responses declined in parallel between 2010 and 2020, which suggests that there was something happening during this time that caused people in these groups to change their minds. In the next section, we will see evidence for a stronger period effect in attitudes about homosexuality.

THE REMARKABLE DECLINE OF HOMOPHOBIA

The GSS includes four questions related to sexual orientation:

1. What about sexual relations between two adults of the same sex—do you think it is always wrong, almost always wrong, wrong only sometimes, or not wrong at all?
2. And what about a man who admits that he is a homosexual? Should such a person be allowed to teach in a college or university, or not?
3. If some people in your community suggested that a book he wrote in favor of homosexuality should be taken out of your public library, would you favor removing this book, or not?
4. Suppose this admitted homosexual wanted to make a speech in your community. Should he be allowed to speak, or not?

If the wording of these questions seems dated, remember that they were written around 1970, when one might "admit" to homosexuality and a large majority thought it was wrong, wrong, or wrong. In general, the GSS avoids changing the wording of questions, because even subtle changes can influence the results. But the price of this consistency is that a phrasing that might have been neutral in 1970 seems loaded today.

Nevertheless, let's look at the results. The following figure shows the percentage of people who chose a homophobic response to these questions as a function of age.

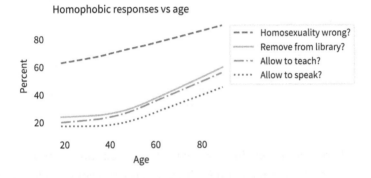

It comes as no surprise that older people are more likely to hold homophobic beliefs. But that doesn't mean people adopt these attitudes as they age. In fact, within every birth cohort, they become less homophobic with age. The following figure shows the results from the first question, showing the percentage of respondents who said homosexuality was wrong (with or without an adverb).

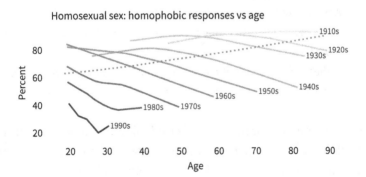

There is clearly a cohort effect: each generation is substantially less homophobic than the one before. And in almost every cohort, homophobia declines with age. But that doesn't mean there is an age effect; if there were, we would expect to see a change in all cohorts at about the same age. And there's no sign of that. So let's see if it might be a period effect. The following figure shows the same results plotted over time rather than age.

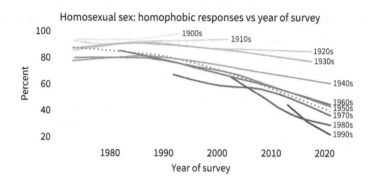

If there is a period effect, we expect to see an inflection point in all cohorts at the same point in time. And there is some evidence of that. Reading from top to bottom, we can see:

- More than 90% of people born in the 1900s and 1910s thought homosexuality was wrong, and they went to their graves without changing their minds.
- People born in the 1920s and 1930s might have softened their views, slightly, starting around 1990.
- Among people born in the 1940s and 1950s, there is a notable inflection point: before 1990, they were almost unchanged; after 1990, they became more tolerant over time.
- In the last four cohorts, there is a clear trend over time, but we did not observe these groups sufficiently before 1990 to identify an inflection point.

On the whole, this looks like a period effect. Also, looking at the overall trend, it declined slowly before 1990 and much more quickly thereafter. So we might wonder what happened in 1990.

WHAT HAPPENED IN 1990?

In general, questions like this are hard to answer. Societal changes are the result of interactions between many causes and effects. But in this case, I think there is an explanation that is at least plausible: advocacy for acceptance of homosexuality has been successful at changing people's minds.

In 1989, Marshall Kirk and Hunter Madsen published a book called *After the Ball* with the prophetic subtitle *How America Will Conquer Its Fear and Hatred of Gays in the '90s*. The authors, with backgrounds in psychology and advertising, outlined a strategy for changing beliefs about homosexuality, which I will paraphrase in two parts: make homosexuality visible, and make it boring. Toward the first goal, they encouraged people to come out and acknowledge their sexual orientation publicly. Toward the second, they proposed a media campaign to depict homosexuality as ordinary.

Some conservative opponents of gay rights latched onto this book as a textbook of propaganda and the written form of the "gay agenda." Of course reality was more complicated than that: social change is the result of many people in many places, not a centrally organized conspiracy.

It's not clear whether Kirk and Madsen's book *caused* America to conquer its fear in the 1990s, but what they proposed turned out to be a remarkable prediction of what happened. Among many milestones, the first National Coming Out Day was celebrated in 1988; the first Gay Pride Day Parade was in 1994 (although previous similar events had used different names); and in 1999, President Bill Clinton proclaimed June as Gay and Lesbian Pride month.

During this time, the number of people who came out to their friends and family grew exponentially, along with the number of openly gay public figures and the representation of gay characters on television and in movies. And as surveys by the Pew Research Center have shown repeatedly, "familiarity is closely linked to tolerance." People who have a gay friend or family member—and know it—are substantially more likely to hold positive attitudes about homosexuality and to support gay rights. All of this adds up to a large period effect that has changed hearts and minds, especially among the most recent birth cohorts.

COHORT OR PERIOD EFFECT?

Since 1990, attitudes about homosexuality have changed due to

- a cohort effect: as old homophobes die, they are replaced by a more tolerant generation; and
- a period effect: within most cohorts, people became more tolerant over time.

These effects are additive, so the overall trend is steeper than the trend within the cohorts—like Simpson's paradox in reverse. But that raises a question: How much of the overall trend is due to the cohort effect, and how much to the period effect?

To answer that, I used a model that estimates the contributions of the two effects separately (a logistic regression model, if you want the details). Then I used the model to generate predictions for two counterfactual scenarios: What if there had been no cohort effect, and what if there had been no period effect? The following figure shows the results.

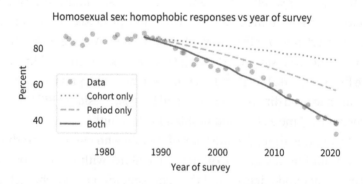

Homosexual sex: homophobic responses vs year of survey

The circles show the actual data. The solid line shows the results from the model from 1987 to 2018, including both effects. The model plots a smooth course through the data, which confirms that it captures the overall trend during this interval. The total change is about 46 percentage points. The dotted line shows what would have happened, according to the model, if there had been no period effect; the total change due to the cohort effect alone would have been about 12 percentage points. The dashed line shows what would have happened if there had been no cohort effect; the total change due to the period effect alone would have been about 29 percentage points.

You might notice that the sum of 12 and 29 is only 41, not 46. That's not an error; in a model like this, we don't expect percentage points to add up (because it's linear on a logistic scale, not a percentage scale). Nevertheless, we can conclude that the magnitude of the period effect is about twice the magnitude of the cohort effect. In other words, most of the change we've seen since 1987 has been due to changed minds, with the smaller part due to generational replacement.

No one knows that better than the San Francisco Gay Men's Chorus. In July 2021, they performed a song by Tim Rosser and Charlie Sohne with the title, "A Message from the Gay Community." It begins like this:

> To those of you out there who are still working against equal rights, we
>   have a message for you [. . .]
> You think that we'll corrupt your kids, if our agenda goes unchecked.
> Funny, just this once, you're correct.
> We'll convert your children, happens bit by bit;
> Quietly and subtly, and you will barely notice it.

Of course, the reference to the gay "agenda" is tongue-in-cheek, and the threat to "convert your children" is only scary to someone who thinks (wrongly) that gay people can convert straight people to homosexuality and who believes (wrongly) that having a gay child is bad. For everyone else, it is clearly a joke. Then the refrain delivers the punchline:

> We'll convert your children; we'll make them tolerant and fair.

For anyone who still doesn't get it, later verses explain:

> Turning your children into accepting, caring people;
> We'll convert your children; someone's gotta teach them not to hate.
> Your children will care about fairness and justice for others.

And finally,

> Your kids will start converting you; the gay agenda is coming home.
> We'll convert your children, and make an ally of you yet.

The thesis of the song is that advocacy can change minds, especially among young people. Those changed minds create an environment where the next generation is more likely to be "tolerant and fair" and where some older people change their minds, too. The data show that this thesis is, "just this once, correct."

THE OVERTON WINDOW

In this chapter, we have considered three questions related to racism, three related to feminism, and four related to homosexuality. For each question I characterized one or more responses as racist, sexist, or homophobic. I realize that these terms are loaded with value judgments, but I think they are accurate. The view that interracial marriage should be illegal is racist. The view that women are less suited to politics than men is sexist. And the view that an "admitted" homosexual should not be allowed to teach at a university is homophobic. Now, so we can put the results on the same y-axis, I will combine these responses into a group I will characterize as "bigoted."

The following figure shows how the percentage of people choosing bigoted responses, for each of the 10 questions we've looked at, has changed over time.

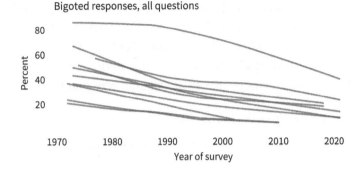

Bigoted responses, all questions

I didn't label the lines because it doesn't matter for now which is which; the point is that they are all headed in the direction of decreasing bigotry. I have smoothed the data, so we don't see short-term variation from year to year. But the long-term trends are consistent; some speed up or slow down, but there are no long-term reversals.

When these questions were first asked in the 1970s, most of the percentages were between 20% and 60%. That's not a coincidence; the topics and the wording of the questions were chosen to avoid very high and very low percentages, which are hard to measure accurately and less informative if you do. There is not much point in asking whether murder is bad. However, many of the views that were mainstream in the 1970s had become unpopular, or on the fringe, by the 2010s. Three of the 10 questions were removed from the survey as they approached the territory of the lizard people, and several more are on their way out.

These results illustrate the "Overton window," which is a term from political science that refers to the range of topics that are part of mainstream discourse. At any point in time, ideas considered acceptable, sensible, or popular are inside the window; ideas considered radical or unthinkable are outside it. Successful politicians, according to the theory, are adept at identifying ideas inside the window and avoiding ideas that are outside it.

Importantly, the Overton window can move over time. On some topics, like the ones we've seen in this chapter, it moves consistently in one direction. On other topics, it has moved back and forth; as an example, consider alcohol prohibition in the United States, which was outside the Overton window prior to the 1870s, became federal policy in 1920, then left the window again after 1933. In the next chapter, we'll use the ideas in this chapter—age-period-cohort analysis and the Overton window—to explain a paradox of political identity.

SOURCES AND RELATED READING

- Planck's autobiography is called *Scientific Autobiography and Other Papers* [98].
- William Lecky described the expanding moral circle in *History of European Morals* [63]. The philosopher Peter Singer wrote about it in *The Expanding Circle* [113], as did Steven Pinker in *The Better Angels of Our Nature* [97]. More recently, Sigal Samuel wrote an article on the topic for *Vox* [108].
- The General Social Survey (GSS) is a project of the independent research organization NORC at the University of Chicago, with principal funding from the National Science Foundation. The data is available from the GSS website [44].
- *Vox* published an article about lizard people [4]. Results from the lizard people poll are reported on the website of Public Policy Polling [30].
- The Pew Research study showing that familiarity breeds acceptance is "Four-in-Ten Americans Have Close Friends or Relatives Who Are Gay" [106].
- You can see a performance of "A Message from the Gay Community" on YouTube [3].
- *Politico* wrote about the origins of the Overton window and its use in popular culture [103]. The example of alcohol prohibition is from the video "The Overton Window of Political Possibility Explained" [122].

CHAPTER 12

CHASING THE OVERTON WINDOW

There's a famous (but unattributable) quote about liberals and conservatives that goes something like this: "If you are not a liberal at 25, you have no heart. If you are not a conservative at 35, you have no brain." It turns out that there is some truth to this. People under 30 are actually more likely to consider themselves liberal, and people over 30 are more likely to consider themselves conservative. And it's not just what they call themselves; the views of older people are more conservative, on average, than the views of younger people.

However, as we saw in the previous chapter, just because older people hold a particular view, that doesn't mean people adopt that view as they get older. In fact, in most birth cohorts, people get more liberal as they age. As you might recognize by now, this is an example of Simpson's paradox. But Simpson's paradox doesn't explain why people *consider* themselves more conservative even as they are becoming more liberal. To understand that, we'll need two ideas from the previous chapter: age-period-cohort analysis and the Overton window.

OLD CONSERVATIVES, YOUNG LIBERALS?

To recap, here are the phenomena I will try to explain:

- People are more likely to identify as conservative as they get older.
- And older people hold more conservative views, on average.

- However, as people get older, their views do not become more conservative.

In other words, people think they are becoming more conservative, but they are not. First, let me show that each of these claims is true; then I'll explain why.

To demonstrate the first claim, I'll use responses to the following question: "I'm going to show you a seven-point scale on which the political views that people might hold are arranged from 'extremely liberal'—point one—to 'extremely conservative'—point seven. Where would you place yourself on this scale?" The points on the scale are "extremely liberal," "liberal," "slightly liberal," "moderate," "slightly conservative," "conservative," and "extremely conservative."

I'll lump the first three points into "liberal" and the last three into "conservative," which makes the number of groups manageable and, it turns out, roughly equal in size. The following figure shows the percentage of respondents who consider themselves to be in each of these groups, as a function of age.

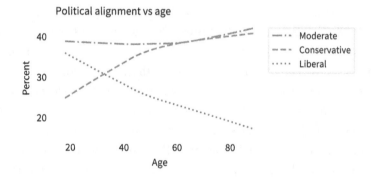

At every age, the largest group of respondents describe themselves as moderate. However, in accordance with the unattributable quip, people under 30 are more likely to consider themselves liberal, and people over 30 are more likely to consider themselves conservative.

Within most birth cohorts, the pattern is the same: as people get older, they are more likely to identify as conservative. The following figure shows the percentage of people who identify as conservative as a function of age, grouped by decade of birth.

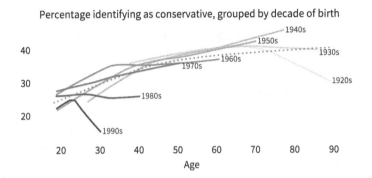

Percentage identifying as conservative, grouped by decade of birth

There is evidence here for an age effect; in many groups there is a notable increase around age 30 and a leveling off after age 40. The most recent cohorts in the dataset, people born in the 1980s and 1990s, might be an exception; so far, they show no trend toward conservative identity. But the oldest of them were surveyed in their 30s, so it might be too early to tell.

WHAT DOES "CONSERVATIVE" MEAN?

Now, to test the second claim—that older people hold more conservative views—we have to figure out what conservative views are. To do that, I searched for questions in the GSS with the biggest differences between the responses of self-described liberals and conservatives. From those, I curated a list of 15 questions that cover a diverse set of topics, with preference for questions that were asked most frequently over the years of the survey.

The topics that made the list are not surprising. They include economic issues like public spending on welfare and the environment; policy issues like the legality of guns, drugs, and pornography; as well as questions related to sex education and prayer in schools, capital punishment, assisted suicide, and (of course) abortion. The list also includes three of the questions we looked at in the previous chapter, related to open housing laws, women in politics, and homosexuality. For the current purpose, we are not concerned with the wording of the questions; we only need to know that liberals and conservatives give different answers. But if you are curious, the list of questions is at the end of this chapter.

For each question, I identified the responses more likely to be chosen by conservatives. The following figure shows, for each question, the percentage of liberals, moderates, and conservatives who chose one of the conservative responses.

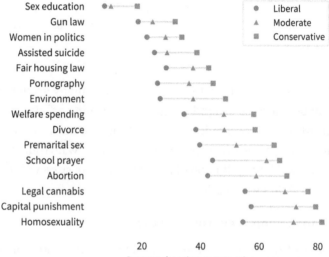

Not surprisingly, conservatives are more likely than liberals to choose conservative responses, and moderates are somewhere in the middle. The differences between the groups range from 11 to 27 percentage points. These results include respondents over the interval from 1973 to 2021, so they are not a snapshot of recent conditions. Rather, they show the issues that have distinguished conservatives from liberals over the past 50 years.

Now let's see if it's true that older people hold more conservative views. For each respondent, I computed the number of conservative responses they chose. One difficulty is that not every respondent answered every question. Some were surveyed before a particular question was added to the survey or after it was removed. Also, in some years respondents were randomly assigned a "ballot" with a subset of the questions. Finally, a small percentage of respondents refuse to answer some questions, or say "I don't know."

To address this difficulty, I used a statistical model to estimate a degree of conservatism for each respondent and to predict the number of conservative responses they would give if they were asked all 15 questions. To avoid extrapolating too far beyond the data, I selected only respondents who answered at least 5 of the 15 questions, which is about 63,000 out of 68,000. The following figure shows the average of these estimates, grouped by political alignment (conservative, moderate, or liberal) and plotted by age.

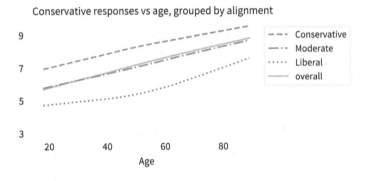

The solid line shows the overall trend: older people hold more conservative views. At age 18, the average number of conservative responses is 5.7 out of 15; at age 89, it is 8.7. The lines for conservatives, moderates, and liberals show that the trend within the groups is the same as the trend between the groups.

However, as we have seen before, that doesn't mean that people get more conservative as they get older. The following figure shows conservatism as a function of age, again, but with respondents grouped by decade of birth.

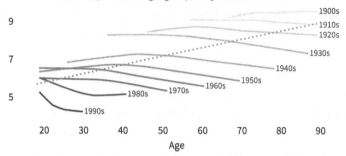

Conservative responses vs age, grouped by decade of birth

Almost every birth cohort became more liberal with age. Some of the oldest cohorts were essentially unchanged. None became more conservative. In several cohorts we see a similar pattern—a relatively fast increase followed by a slower decrease—but the inflection point does not appear at the same age in different groups, so that does not suggest an age effect.

HOW CAN THIS BE?

By now you have probably figured out that part of the explanation is Simpson's paradox. Older respondents are more likely to come from earlier generations, which were more conservative; younger respondents are more likely to come from recent generations, which are more liberal. That explains why older people are more conservative, but people don't become more conservative as they age. The apparent increase is a cohort effect, not an age effect.

But that doesn't explain why people are more likely to *consider* themselves conservative even as, on average, they become more liberal. The explanation I propose has three pieces:

- First, the center of public opinion has moved toward liberalism over the past 50 years, primarily due to a cohort effect and secondarily due to a period effect.
- Second, "liberal" and "conservative" are relative terms. If someone's views are left of center, they are more likely to call themselves liberal. If their views are right of center, they are more likely to call themselves

conservative. Either way, the label they choose depends on where they think the center is.

- Third, over the past 50 years, both liberals and conservatives have become more liberal. As a result, conservatives are about as liberal now as liberals were in the 1970s.

Putting these pieces together, the consequence is that someone considered liberal in the 1970s would be considered conservative in the 2010s. So someone considered liberal in their youth, and who becomes more liberal over time, might nevertheless be considered—and identify as—more conservative as they age.

Let's consider these points one at a time.

THE CENTER DOES NOT HOLD

In the past 50 years, the Overton window has shifted toward liberalism. The following figure shows how the answers to my 15 questions have changed over time. Each circle represents the actual average for a group of 2000 respondents. The dotted line shows a smooth curve fitted to the data.

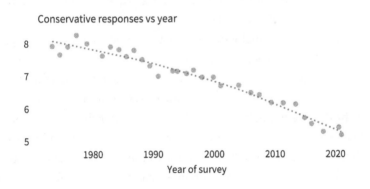

The average number of conservative responses has decreased. In 1973, the average respondent chose a conservative response to 8.1 out of 15 questions; in 2021, it had fallen to 5.3. Also, contrary to what you might have heard, there is no evidence that this trend has slowed or reversed recently. If anything, it might be accelerating. To see whether this trend is driven primarily by period or cohort effects, consider the following graph, which shows conservatism grouped by decade of birth and plotted over time.

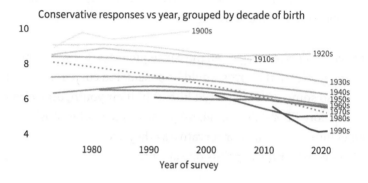

There is clearly a cohort effect: almost every birth cohort is more liberal than the one before it. If there is also period effect, we expect to see all groups moving up and down at the same time. And it looks like they did: several cohorts trended toward conservative before 1990 and toward liberal after. But the net change in most groups is small, so it seems like most of the change we've seen is due to the cohort effect, not the period effect.

To quantify that conclusion, we can run counterfactual models, as we did in the previous chapter with attitudes toward homosexuality. In that example, we found that the cohort effect accounted for about one-third of the change, and the period effect accounted for the other two-thirds. The following figure shows the actual change in conservatism over the past 50 years and two counterfactual models: one with the period effect only and one with the cohort effect only. The cohort effect is about five times bigger than the period effect, which means that the observed changes are due primarily to generational replacement and only secondarily to changed minds.

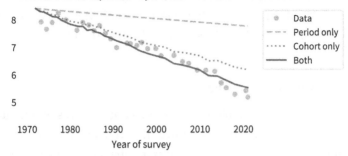

Conservative responses vs year, with counterfactual models

EVERYTHING IS RELATIVE

So again, if people are actually becoming more liberal, why do they think they are getting more conservative? The second part of the explanation is that people classify their political views relative to the perceived center. As evidence, consider the following figure, which shows the percentage of people who identify as moderate, conservative, or liberal over time.

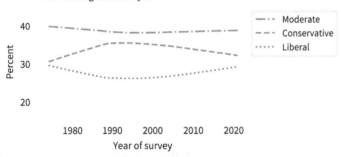

Political alignment vs year

The percentage of people who identify as "Moderate" has barely changed in 50 years. During the 1980s, the percentage of conservatives increased and the percentage of liberals declined. I conjecture that these changes are the result of negative connotations that were assigned to the word "liberal" during the Reagan Era. Many left-of-center politicians avoided using the word for decades; GSS respondents might have been reluctant to identify with the word, even if their views were actually left of center.

Despite all that, the responses are remarkably stable over time.

These results suggest that the label people choose for themselves depends primarily on how they compare their views to the center of public opinion and maybe secondarily on the connotations of the labels.

ARE WE MORE POLARIZED?

The third part of the explanation is that conservatives, moderates, and liberals have all become more liberal over time. In current commentary, it is often taken for granted that political views in the United States have become more polarized. If we take that to mean that conservatives are becoming more conservative, and liberals more liberal, that turns out not to be true. The following figure shows conservatism over time, grouped by political alignment.

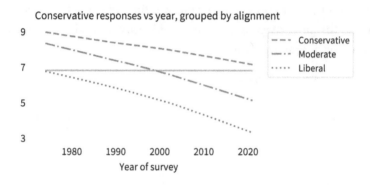

All three groups have become more liberal; however, the slopes of the lines are somewhat different. Over this interval, conservatives have become more liberal by about 1.9 responses, on average, moderates by 3.2, and liberals by 3.4. That's a kind of polarization in the sense that the groups moved farther apart, but not in the sense that they are moving in opposite directions. Also, the magnitude of the divergence is modest. If civil discourse was possible in 1973, when conservatives and liberals disagreed about 2.5 questions, on average, it should be possible now, when they disagree about 3.7.

CHASING OVERTON

Now we are ready to put the pieces of the explanation together. In the previous figure, the horizontal line is at 6.8 responses, which was the expected level of conservatism in 1974 among people who called themselves liberal. Suppose you take a time machine back to 1974, find an average liberal, and bring them to the turn of the millennium. Their responses to the 15 questions indicate that they would be indistinguishable from the average moderate in 2000. And if you bring them to 2021, their quaint 1970s liberalism would be almost as conservative as the average conservative. In 1974, they were left of center and would probably identify as liberal. In 2021, they would find themselves considerably right of center and would probably identify as conservative.

This time machine scenario is almost precisely what happened to people born in the 1940s. The following figure is the same as the previous one, except that the solid line shows average conservatism for people born in the 1940s.

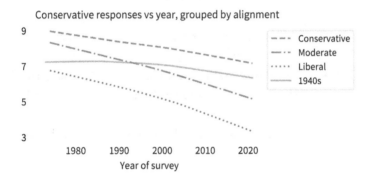

In the 1970s, when they were in their 30s, they were almost as liberal, on average, as the people who identified as liberal. Over their lives, they trended toward conservative until 1990, reversed toward liberal after, and ended up a little left of where they started. Meanwhile, the world around them changed. In the 1990s, when they were in their 50s, they became indistinguishable from the people around them who identified as moderate. And in the 2010s, when they were in their 70s, they found—to their surprise and often dismay—that they were almost as conservative, on average, as people who identified as conservative.

Abe Simpson, the grandfather on *The Simpsons*, summed up their plight: "I used to be with 'it,' but then they changed what 'it' was. Now what I'm with isn't 'it' and what's 'it' seems weird and scary to me. It'll happen to you!" He's right. Unless your views move faster than the Overton window—and for most people they don't—you will go to bed one night thinking you're with "it," and the next day you'll wake up on the wrong side of history.

SOURCES AND RELATED READING

In many chapters, I started with analysis from a published article and then replicated or extended it. This chapter is a little different because it is mostly about my own investigation using data from the General Social Survey. As a result, I have fewer sources to cite and less related reading to recommend. Also, it would be appropriate to consider the results in this chapter an initial exploration by a data scientist, intended to demonstrate certain tools and concepts, not definitive research by experts in political science.

- Quote Investigator explores the origin of the unattributable quote [88].
- The General Social Survey (GSS) is a project of the independent research organization NORC at the University of Chicago, with principal funding from the National Science Foundation. The data is available from the GSS website [44].
- An article in the *Chicago Tribune* discusses social attitudes toward the word "liberal" [89].

APPENDIX: THE 15 QUESTIONS

This appendix provides the wording of the 15 questions from the General Social Survey that most distinguish liberals and conservatives, identified by topic and the GSS code name.

- Homosexuality (homosex): What about sexual relations between two adults of the same sex—do you think it is always wrong, almost always wrong, wrong only sometimes, or not wrong at all?
- Capital punishment (cappun): Do you favor or oppose the death penalty for persons convicted of murder?
- Legal cannabis (grass): Do you think the use of marijuana should be made legal or not?

- Abortion (abany): Please tell me whether or not you think it should be possible for a pregnant woman to obtain a legal abortion if the woman wants it for any reason?

- Prayer in public schools (prayer): The United States Supreme Court has ruled that no state or local government may require the reading of the Lord's Prayer or Bible verses in public schools. What are your views on this—do you approve or disapprove of the court ruling?

- Premarital sex (premarsx): There's been a lot of discussion about the way morals and attitudes about sex are changing in this country. If a man and woman have sex relations before marriage, do you think it is always wrong, almost always wrong, wrong only sometimes, or not wrong at all?

- Divorce (divlaw): Should divorce in this country be easier or more difficult to obtain than it is now?

- Spending on welfare (natfare) and the environment (natenvir): We are faced with many problems in this country, none of which can be solved easily or inexpensively. I'm going to name some of these problems, and for each one I'd like you to tell me whether you think we're spending too much money on it, too little money, or about the right amount.

- Pornography (pornlaw): Which of these statements comes closest to your feelings about pornography laws?
  - There should be laws against the distribution of pornography whatever the age.
  - There should be laws against the distribution of pornography to persons under 18.
  - There should be no laws forbidding the distribution of pornography.

- Open housing law (racopen): Suppose there is a community-wide vote on the general housing issue. There are two possible laws to vote on; which law would you vote for?
  - One law says that a homeowner can decide for himself whom to sell his house to, even if he prefers not to sell to [people of a particular race].
  - The second law says that a homeowner cannot refuse to sell to someone because of their race or color.

- Assisted suicide (letdie1): When a person has a disease that cannot be cured, do you think Doctors should be allowed by law to end the patient's life by some painless means if the patient and his family request it?

- Women in politics (fepol): Tell me if you agree or disagree with this statement: Most men are better suited emotionally for politics than are most women.
- Gun control (gunlaw): Would you favor or oppose a law which would require a person to obtain a police permit before he or she could buy a gun?
- Sex education (sexeduc): Would you be for or against sex education in the public schools?

EPILOGUE

I started this book with the premise that we are better off when our decisions are guided by evidence and reason. But I never said it would be easy. In practice, data can behave in ways that violate our expectations and intuition—so much that we've nearly worn out the word "paradox."

And it's easy to be fooled. In the previous chapters we've seen several ways things can go wrong. One is selection bias, like length-biased sampling in chapter 2 and collision bias in chapter 7. Other ways to be fooled include combining groups that should be considered separately, as in chapter 10, and mistaking a period effect for an age effect, as in chapter 11. These errors can distort not just our measurements, like the sizes of classes, but also our perceptions of ourselves, like feeling normal or not, and our perceptions of the world, like the fairness of the criminal justice system or the efficacy of vaccines.

But working with data might not be as tricky as I've made it seem, because this book also suffers from selection bias. I chose the examples I thought were the most interesting and relevant to real-world problems, but as it turns out, some of them are also the most confusing and counterintuitive. Usually it's not *that* hard; we really can use data to resolve questions and settle arguments.

To make that possible, we need three things that go together: questions, data, and methods. For most projects, the hardest part is to find the data that can answer the questions you care about. If you can do that, the methods you need are often as simple as counting and comparing. And good data visualization goes a long way. Here's one of my favorite examples. When my wife and I were expecting our first child, we heard that first babies tend to be born early. We also heard that first babies tend to be born late. So, which is it?

Being who I am, I went looking for data and quickly found the National Survey of Family Growth (NSFG), which is run by the National Center for Health Statistics. Since 1973, they have surveyed a representative sample of adults in the United States and gathered "information on family life, marriage and divorce, pregnancy, infertility, use of contraception, and men's and women's health." Their datasets are free to download from their web page. You might remember that we used birthweights from the NSFG in chapter 4 and pregnancy durations in chapter 5.

Using this data, I found a sample of first babies and others and compared their lengths of gestation. As it turns out, first babies are slightly more likely than others to be born early or late and less likely to be born at the nominal gestation length of 39 weeks. On average, they are born about 13 hours later. I don't have a biological explanation for that difference, but it is consistent over time, and it persists even if we control for medical interventions like induced labor and cesarean delivery.

So I wrote a blog article about it, which I posted in February 2011. It was only the sixth article on my new blog, which I called "Probably Overthinking It." Since then, it has always been my most-read article; as of October 2022, it has more than 210,000 page views.

Now, if you have a web browser nearby, open the search engine of your choice and type in "first babies early or late." The chances are that my article, or one of the follow-ups to it, is near the top of the page. And if you click on the others, you'll find that a lot of them reference my article. Somehow, I am now recognized as the go-to expert on this topic.

I'm not telling this story to brag; my point is that what I did required nothing special. I asked a question and found data that could answer it; I used basic statistics to analyze the data and created simple visualizations to show the results; and then I wrote an article and put it on the internet. Maybe not anyone could have done it, but a lot of people could.

Data and the tools to work with it are more accessible now than ever. Obviously, the internet is a big part of that, along with the services that make information easy to find. The other big part is the emerging ethos of open, reproducible science. Not that long ago, like when I was in graduate school, there was no expectation that researchers would make their data or computer code available. If you wanted to replicate someone's experiment or run a new analysis on their data, you had to contact them directly, with the understanding that the data might not be available and you might not even get a reply. And if you wanted to run someone else's code, good luck. It might not be available, or they might not be willing to share. The quality of the code was likely to be poor, with little or no documentation. And it might not be possible to replicate the environment where the code ran.

Things now are not perfect, but they are a lot better. Working on the examples in this book, one of my strategies was to replicate previous work and then extend it. Sometimes I updated it with new data, sometimes I applied different methods, and sometimes I took it too far, like the immortal Swede in chapter 5.

In most cases, my attempts to replicate prior work went well. Often I was able to download data from primary sources, many of them government agencies. Other times the authors made the data available on the web, either on their own site or their publisher's. And when I had to contact people to ask for data, many of them replied quickly and helpfully. There were only a few examples where I had to resort to shenanigans like extracting data from a digital figure, and most of those were from papers published before widespread use of computers.

Tools and practices for sharing code have not come as far, but they are improving. Toward that end, I have made all of the code for this book available online [35]. I hope that at least some readers will take the chance to replicate and extend what I have done. I look forward to seeing what they find.

ACKNOWLEDGMENTS

Thank you to my editor at University of Chicago Press, Joseph Calamia, and the other people at UCP who worked on this book: Matt Lang in acquisitions, Tamara Ghattas in manuscript editing, Brian Chartier in design, Skye Agnew and Annika Rae in production, and Anne Strother in promotions. Special thanks to the reviewers and technical reviewers.

Thanks to my former employer, Olin College of Engineering, for their support of this project from the earliest days of the "Probably Overthinking It" blog, and to my colleagues, Professors Jonathan Adler, Sara Hendren, and Carrie Nugent for their comments and suggestions.

Thanks to my current employer, DrivenData, for their ongoing support, and thanks to my colleagues who read chapters and gave helpful feedback on my practice talk.

Whenever possible, I have sent draft chapters to relevant experts for comments. I appreciate the time they took and the suggestions they made. They include student of data science Kayla Brand; Professor Ryan Burge at Eastern Illinois University; Professor Aaron Clauset at the University of Colorado Boulder; Professor Dennis Culhane at the University of Pennsylvania; data scientist Cameron Davidson-Pilon, author of the lifelines package for survival analysis; Professor Allison Horst at the University of California, Santa Barbara; Dr. Frietson Galis at the Naturalis Biodiversity Center; Philip Gingerich, Professor Emeritus at the University of Michigan; Professor Sonia Hernández-Díaz at Harvard University; Dr. Nidhi Kalra and Dr. Shawn Bushway at the RAND Corporation; Professor Jeffrey Morris at the University of Pennsylvania; Professor Sergey Nigai at the University of Colorado Boulder; Floyd Norris, formerly Chief Financial Correspondent of the *New York Times* and now at Johns Hopkins University; NOAA research scientist Dr. Courtney Peck; Samuel Preston, Professor at the University of Pennsylvania; Dr. Adrian Price-Whelan at the Flatiron Institute; Todd Rose, author of *The End of Average*; data scientist Ethan Rosenthal; scientist and statistician Dr. Patrick Royston; Professor Enrique Schisterman at the University of Pennsylvania; economist Dr. Andrea Stella at The Federal Reserve; Dr. Francisco Villavicencio at Johns Hopkins University; research scientist at the Planetary Science Institute, Dr. Kat Volk. Of course none of them are responsible for my errors.

Special thanks to people who read several early chapters, including June Downey and Dr. Jennifer Tirnauer.

And extra special thanks to the child who was one week early and the child who was two weeks late for inspiring the article that got the whole thing started, and to Lisa Downey for her love and support throughout this project, not to mention copyediting every chapter.

Thank you!

BIBLIOGRAPHY

[1] "Distribution of Women Age 40–50 by Number of Children Ever Born and Marital Status: CPS, Selected Years, 1976–2018." United States Census Bureau, 2018. https://www.census.gov/data/tables/time-series/demo/fertility/his-cps.html.

[2] "27th Anniversary Edition James Joyce Ramble 10K." Cool Running, 2010. https://web.archive.org/web/20100429073703/http://coolrunning.com/results/10/ma/Apr25_27thAn_set1.shtml.

[3] "A Message from the Gay Community." San Francisco Gay Men's Chorus, YouTube, July 1, 2021. https://www.youtube.com/watch?v=ArOQF4kadHA.

[4] Alex Abad-Santos. "Lizard People: The Greatest Political Conspiracy Ever Created." Vox, February 20, 2015. https://www.vox.com/2014/11/5/7158371/lizard-people-conspiracy-theory-explainer.

[5] Gregor Aisch and Amanda Cox. "A 3-D View of a Chart That Predicts the Economic Future: The Yield Curve." New York Times, March 19, 2015. https://www.nytimes.com/interactive/2015/03/19/upshot/3d-yield-curve-economic-growth.html.

[6] Robin George Andrews. "How a Tiny Asteroid Strike May Save Earthlings from City-Killing Space Rocks." New York Times, March 21, 2022. https://www.nytimes.com/2022/03/21/science/nasa-asteroid-strike.html.

[7] Julia Angwin, Jeff Larson, Surya Mattu, and Lauren Kirchner. "Machine Bias: There's Software Used across the Country to Predict Future Criminals. And It's Biased Against Blacks." ProPublica, May 23, 2016. https://www.propublica.org/article/machine-bias-risk-assessments-in-criminal-sentencing.

[8] "Anthropometric Survey of US Army Personnel." US Army Natick Soldier Research, Development and Engineering Center. 2012. https://www.openlab.psu.edu/ansur2/.

[9] "Archive of Wednesday, 7 September 2005." SpaceWeatherLive.com, 2005. https://www.spaceweatherlive.com/en/archive/2005/09/07/xray.html.

[10] Hailey R. Banack and Jay S. Kaufman. "The "Obesity Paradox" Explained." Epidemiology 24.3 (2013): 461–62.

[11] "Behavioral Risk Factor Surveillance System Survey Data." Centers for Disease Control and Prevention (CDC), 2020. https://www.cdc.gov/brfss.

[12] Hiram Beltrán-Sánchez, Eileen M. Crimmins, and Caleb E. Finch. "Early Cohort Mortality Predicts the Rate of Aging in the Cohort: A Historical Analysis." *Journal of Developmental Origins of Health and Disease* 3.5 (2012): 380–86.

[13] *Big Five Personality Test.* Open-Source Psychometrics Project. 2019. https://openpsychometrics.org/tests/IPIP-BFFM.

[14] George E. Bigelow, Warren E. Bickel, John D. Roache, Ira A. Liebson, and Pat Nowowieski. "Identifying Types of Drug Intoxication: Laboratory Evaluation of a Subject-Examination Procedure." Tech. rep. DOT HS 806 753. National Highway Traffic Safety Administration, 1985. http://www.decp.us/pdfs/Bigelow_1985_DRE_validation_study.pdf.

[15] "Carrington Event." Wikipedia, 2022. https://en.wikipedia.org/wiki/Carrington_Event.

[16] C. R. Charig, D. R. Webb, S. R. Payne, and J. E. A. Wickham. "Comparison of Treatment of Renal Calculi by Open Surgery, Percutaneous Nephrolithotomy, and Extracorporeal Shockwave Lithotripsy." *British Medical Journal (Clinical Research Edition)* 292.6524 (1986): 879–82.

[17] Y. B. Cheung, P. S. F. Yip, and J. P. E. Karlberg. "Mortality of Twins and Singletons by Gestational Age: A Varying-Coefficient Approach." *American Journal of Epidemiology* 152.12 (2000): 1107–16.

[18] "Child Mortality Rate, under Age Five." Gapminder Foundation. 2022. https://www.gapminder.org/data/documentation/gd005/.

[19] "China Is Trying to Get People to Have More Babies." *The Economist*, September 29, 2022. https://www.economist.com/china/2022/09/29/china-is-trying-to-get-people-to-have-more-babies.

[20] Richard P. Compton. "Field Evaluation of the Los Angeles Police Department Drug Detection Program." Tech. rep. DOT HS 807 012, 1986. https://trid.trb.org/view/466854.

[21] John D. Cook. "Student-t as a Mixture of Normals." 2009. https://www.johndcook.com/blog/2009/10/30/student-t-mixture-normals/.

[22] Sam Corbett-Davies, Emma Pierson, Avi Feller, and Sharad Goel. "A Computer Program Used for Bail and Sentencing Decisions Was Labeled Biased against Blacks. It's Actually Not That Clear." *Washington Post*, October 17, 2016. https://www.washingtonpost.com/news/monkey-cage/wp/2016/10/17/can-an-algorithm-be-racist-our-analysis-is-more-cautious-than-propublicas/.

[23] "Could Solar Storms Destroy Civilization? Solar Flares & Coronal Mass Ejections." Kurzgesagt—In a Nutshell. YouTube, June 7, 2020. https://www.youtube.com/watch?v=oHHSSJDJ4oo.

[24] "COVID-19 Vaccine Surveillance Report, Week 38." Public Health England, September 23, 2021. https://assets.publishing.service.gov.uk/government/uploads/system/uploads/attachment_data/file/1019992/Vaccine_surveillance_report_-_week_38.pdf.

[25] Tyler Cowen. "Six Rules for Dining Out: How a Frugal Economist Finds the

Perfect Lunch." *The Atlantic*, May 2012. https://www.theatlantic.com/magazine /archive/2012/05/six-rules-for-dining-out/308929/.

[26] Gilbert S. Daniels. "The 'Average Man'?" Tech. rep. RDO No. 695-71. Air Force Aerospace Medical Research Lab, Wright-Patterson AFB, Ohio, 1952. https://apps .dtic.mil/sti/pdfs/AD0010203.pdf.

[27] D. Scott Davis, David A. Briscoe, Craig T. Markowski, Samuel E. Saville, and Christopher J. Taylor. "Physical Characteristics That Predict Vertical Jump Performance in Recreational Male Athletes." *Physical Therapy in Sport* 4.4 (2003): 167–74.

[28] Jill De Ron, Eiko I. Fried, and Sacha Epskamp. "Psychological Networks in Clinical Populations: Investigating the Consequences of Berkson's Bias." *Psychological Medicine* 51.1 (2021): 168–76.

[29] "Deaths by Vaccination Status, England." UK Office for National Statistics, July 6, 2022. https://www.ons.gov.uk/peoplepopulationandcommunity /birthsdeathsandmarriages/deaths/datasets/deathsbyvaccinationstatusengland.

[30] "Democrats and Republicans Differ on Conspiracy Theory Beliefs." Public Policy Polling. April 2, 2013. https://www.publicpolicypolling.com/wp-content /uploads/2017/09/PPP_Release_National_ConspiracyTheories_040213.pdf.

[31] "Birth to 36 Months: Boys; Length-for-Age and Weight-for-Age Percentiles." Centers for Disease Control and Prevention (CDC), 2022. https://www.cdc.gov /growthcharts/data/set1clinical/cj41l017.pdf.

[32] "Distribution of Undergraduate Classes by Course Level and Class Size." Purdue University, 2016. https://web.archive.org/web/20160415011613/https://www .purdue.edu/datadigest/2013-14/InstrStuLIfe/DistUGClasses.html.

[33] Allen B. Downey. *Elements of Data Science: Getting Started with Data Science and Python*. Green Tea Press, 2021.

[34] ———. *Elements of Data Science: Recidivism Case Study*. 2021. https://allendowney .github.io/RecidivismCaseStudy.

[35] ———. *Probably Overthinking It: Online Resources*. 2022. https://allendowney .github.io/ProbablyOverthinkingIt.

[36] *Ebner v. Cobb County*. ACLU. September 25, 2017. https://acluga.org/ebner-v -cobb-county.

[37] Shah Ebrahim. "Yerushalmy and the Problems of Causal Inference." *International Journal of Epidemiology* 43.5 (2014): 1349–51.

[38] Jordan Ellenberg. *How Not to Be Wrong: The Power of Mathematical Thinking*. Penguin, 2015.

[39] *The Joe Rogan Experience* podcast, episode 1717. October 15, 2021. https:// open.spotify.com/episode/1VNcMVzwgdU2gXdbw7yqCL?si=WjK0 _EQ7TXGgFdVmda_gRw.

[40] K. Anders Ericsson. "Training History, Deliberate Practice and Elite Sports Performance: An Analysis in Response to Tucker and Collins Review—What Makes Champions?" *British Journal of Sports Medicine* 47.9 (2013): 533–35.

[41] Scott L. Feld. "Why Your Friends Have More Friends Than You Do." *American Journal of Sociology* 96.6 (1991): 1464–77.

[42] Erwin Fleischmann, Nancy Teal, John Dudley, Warren May, John D. Bower, and Abdulla K. Salahudeen. "Influence of Excess Weight on Mortality and Hospital Stay in 1346 Hemodialysis Patients." *Kidney International* 55.4 (1999): 1560–67.

[43] Frietson Galis, Inke Van Der Sluijs, Tom J. M. Van Dooren, Johan A. J. Metz, and Marc Nussbaumer. "Do Large Dogs Die Young?" *Journal of Experimental Zoology, Part B: Molecular and Developmental Evolution* 308.2 (2007): 119–26.

[44] "General Social Survey." NORC at the University of Chicago, 2022. https://gss .norc.org/Get-The-Data.

[45] Philip D. Gingerich. "Arithmetic or Geometric Normality of Biological Variation: An Empirical Test of Theory." *Journal of Theoretical Biology* 204.2 (2000): 201–21.

[46] Malcolm Gladwell. *Outliers: The Story of Success.* Little, Brown, 2008.

[47] *Global Leaderboard.* Chess.com, March 1, 2022. https://www.chess.com /leaderboard/live.

[48] Harvey Goldstein. "Commentary: Smoking in Pregnancy and Neonatal Mortality." *International Journal of Epidemiology* 43.5 (2014): 1366–68.

[49] Kristen B. Gorman, Tony D. Williams, and William R. Fraser. "Ecological Sexual Dimorphism and Environmental Variability within a Community of Antarctic Penguins (Genus Pygoscelis)." *PLOS One* 9.3 (2014): e90081.

[50] Gareth J. Griffith, Tim T. Morris, Matthew J. Tudball, Annie Herbert, Giulia Mancano, Lindsey Pike, et al. "Collider Bias Undermines Our Understanding of COVID-19 Disease Risk and Severity." *Nature Communications* 11.1 (2020): 1–12.

[51] Sara Hendren. *What Can a Body Do? How We Meet the Built World.* Penguin, 2020.

[52] Sonia Hernández-Díaz, Enrique F. Schisterman, and Miguel A. Hernán. "The Birth Weight 'Paradox' Uncovered?" *American Journal of Epidemiology* 164.11 (2006): 1115–20.

[53] Allison Horst. "Example Graphs Using the Penguins Data." Palmerpenguins, n.d. https://allisonhorst.github.io/palmerpenguins/articles/examples.html.

[54] Jennifer Hotzman, Claire C. Gordon, Bruce Bradtmiller, Brian D. Corner, Michael Mucher, Shirley Kristensen, et al. "Measurer's Handbook: US Army and Marine Corps Anthropometric Surveys, 2010–2011." Tech. rep. US Army Natick Soldier Research, Development and Engineering Center, 2011.

[55] Steven Johnson. *Extra Life: A Short History of Living Longer.* Penguin, 2022.

[56] *JPL Small-Body Database.* Jet Propulsion Laboratory, August 15, 2018. https://ssd .jpl.nasa.gov/tools/sbdb_lookup.html.

[57] D. Kahneman, O. Sibony, and C. R. Sunstein. *Noise: A Flaw in Human Judgment.* HarperCollins, 2021.

[58] Brendan Keefe and Michael King. "The Drug Whisperer: Drivers Arrested while Stone Cold Sober." *11 Alive,* WXIA-TV, January 31, 2018. https://www.11alive.com /article/news/investigations/the-drug-whisperer-drivers-arrested-while-stone -cold-sober/85-502132144.

[59] Katherine M. Keyes, George Davey Smith, and Ezra Susser. "Commentary: Smoking in Pregnancy and Offspring Health: Early Insights into Family-Based

and 'Negative Control' Studies?" *International Journal of Epidemiology* 43.5 (2014): 1381–88.

[60] Sarah Kliff and Aatish Bhatia. *When They Warn of Rare Disorders, These Prenatal Tests Are Usually Wrong.* New York Times, January. 1, 2022. https://www.nytimes .com/2022/01/01/upshot/pregnancy-birth-genetic-testing.html.

[61] Gina Kolata. "Blood Tests That Detect Cancers Create Risks for Those Who Use Them." *New York Times,* June 10, 2022. https://www.nytimes.com/2022/06/10 /health/cancer-blood-tests.html.

[62] Michael S. Kramer, Xun Zhang, and Robert W. Platt. "Commentary: Yerushalmy, Maternal Cigarette Smoking and the Perinatal Mortality Crossover Paradox." *International Journal of Epidemiology* 43.5 (2014): 1378–81.

[63] William Edward Hartpole Lecky. *History of European Morals, from Augustus to Charlemagne.* Vol. 1. D. Appleton, 1897. https://www.gutenberg.org/files/39273 /39273-h/39273-h.html.

[64] Ellen Lee. "At-Home COVID-19 Antigen Test Kits: Where to Buy and What You Should Know." *New York Times,* December 21, 2021. https://www.nytimes.com /wirecutter/reviews/at-home-covid-test-kits.

[65] Daniel J. Levitin. *This Is Your Brain on Music: The Science of a Human Obsession.* Penguin, 2006.

[66] Dyani Lewis. "Why Many Countries Failed at COVID Contact-Tracing—but Some Got It Right." *Nature* 588.7838 (2020): 384–88.

[67] Michael Lewis. *The Premonition: A Pandemic Story.* Penguin UK, 2021.

[68] Rolv T. Lie. "Invited Commentary: Intersecting Perinatal Mortality Curves by Gestational Age—Are Appearances Deceiving?" *American Journal of Epidemiology* 152.12 (2000): 1117–19.

[69] "Life Tables by Country." WHO Global Health Observatory, 2022. https://apps .who.int/gho/data/node.main.LIFECOUNTRY?lang=en.

[70] "List of Disasters by Cost." Wikipedia, 2022. https://en.wikipedia.org/wiki/List _of_disasters_by_cost.

[71] David Lusseau, Karsten Schneider, Oliver J. Boisseau, Patti Haase, Elisabeth Slooten, and Steve M. Dawson. "The Bottlenose Dolphin Community of Doubt-ful Sound Features a Large Proportion of Long-Lasting Associations." *Behavioral Ecology and Sociobiology* 54.4 (2003): 396–405.

[72] William MacAskill, Teruji Thomas, and Aron Vallinder. "The Significance, Persistence, Contingency Framework." GPI Technical Report No. T1-2022. Global Priorities Institute, 2022. https://globalprioritiesinstitute.org/wp -content/uploads/William-MacAskill-Teruji-Thomas-and-Aron-Vallinder-The -Significance-Persistence-Contingency-Framework.pdf.

[73] Benoit B. Mandelbrot. *The Fractal Geometry of Nature.* Vol. 1. Freeman, 1982.

[74] Arjun K. Manrai, Gaurav Bhatia, Judith Strymish, Isaac S. Kohane, and Sachin H. Jain. "Medicine's Uncomfortable Relationship with Math: Calculating Positive Predictive Value." *JAMA Internal Medicine* 174.6 (2014): 991–93.

[75] "MBTA Data." Massachusetts Bay Transportation Authority, 2021. https://www .mbta.com/developers.

[76] Julian J. McAuley and Jure Leskovec. "Learning to Discover Social Circles in Ego Networks." In *Advances in Neural Information Processing Systems 25 (NIPS 2012)*, ed. Peter L. Bartlett et al., 2012, 548–56. http://dblp.uni-trier.de/db/conf /nips/nips2012.html#McAuleyL12.

[77] Richard McElreath. *Statistical Rethinking: A Bayesian Course with Examples in R and Stan*. Chapman and Hall/CRC, 2020.

[78] H. Jay Melosh. *Planetary Surface Processes*. Vol. 13. Cambridge University Press, 2011.

[79] V. J. Menon and D. C. Agrawal. "Renewal Rate of Filament Lamps: Theory and Experiment." *Journal of Failure Analysis and Prevention* 7.6 (2007): 419–23.

[80] Dasia Moore. "State Suspends COVID-19 Testing at Orig3n, Boston Lab Responsible for at Least 383 False Positive Results." *Boston Globe*, September 8, 2020. https://www.bostonglobe.com/2020/09/08/nation/state-suspends -covid-19-testing-orig3n-boston-based-lab-responsible-least-383-false-positive -results/.

[81] Jeffrey Morris. "UK Data: Impact of Vaccines on Deaths." November 27, 2021. https://www.covid-datascience.com/post/what-do-uk-data-say-about-real-world -impact-of-vaccines-on-all-cause-deaths.

[82] Randall Munroe. "Base Rate." *xkcd*, 2021. https://xkcd.com/2476/.

[83] "National Longitudinal Survey of Youth 1997 (NLSY97)." US Bureau of Labor Statistics, 2018. https://www.nlsinfo.org/content/cohorts/nlsy97.

[84] "National Survey of Family Growth." Centers for Disease Control and Prevention (CDC), 2019. https://www.cdc.gov/nchs/nsfg.

[85] Floyd Norris. "Can Every Group Be Worse Than Average? Yes." *New York Times*, May 1, 2013. https://economix.blogs.nytimes.com/2013/05/01/can-every-group -be-worse-than-average-yes.

[86] ———. "Median Pay in U.S. Is Stagnant, but Low-Paid Workers Lose." *New York Times*, April 27, 2013. https://www.nytimes.com/2013/04/27/business/economy /wage-disparity-continues-to-grow.html.

[87] Numberphile. "Does Hollywood Ruin Books?" YouTube, August 28, 2018. https:// www.youtube.com/watch?v=FUD8h9JpEVQ.

[88] Garson O'Toole. "If You Are Not a Liberal at 25, You Have No Heart. If You Are Not a Conservative at 35, You Have No Brain." Quote Investigator, February 24, 2014. https://quoteinvestigator.com/2014/02/24/heart-head.

[89] Clarence Page and a member of the *Chicago Tribune*'s editorial board. "When Did 'Liberal' Become a Dirty Word?" *Chicago Tribune*, July 20, 2007. https://www .chicagotribune.com/news/ct-xpm-2007-07-29-0707280330-story.html.

[90] Lionel Page. "Everybody Should Know about Berkson's Paradox." Twitter, 2021. https://twitter.com/page_eco/status/1373266475230789633.

[91] Mark Parascandola. "Commentary: Smoking, Birthweight and Mortality: Jacob Yerushalmy on Self-Selection and the Pitfalls of Causal Inference." *International Journal of Epidemiology* 43.5 (2014): 1373–77.

[92] John Allen Paulos. "Why You're Probably Less Popular Than Your Friends." *Scientific American* 304.2 (2011): 33.

[93] Judea Pearl and Dana Mackenzie. *The Book of Why: The New Science of Cause and Effect*. Basic Books, 2018.

[94] Caroline Criado Perez. *Invisible Women: Data Bias in a World Designed for Men*. Abrams, 2019.

[95] ———. "The Deadly Truth about a World Built for Men—From Stab Vests to Car Crashes." *The Guardian*, February 23, 2019. https://www.theguardian.com /lifeandstyle/2019/feb/23/truth-world-built-for-men-car-crashes.

[96] "Personal Protective Equipment and Women." Trades Union Congress (TUC), 2017. https://www.tuc.org.uk/research-analysis/reports/personal-protective -equipment-and-women.

[97] Steven Pinker. *The Better Angels of Our Nature: Why Violence Has Declined*. Penguin, 2012.

[98] Max Planck. *Scientific Autobiography and Other Papers*. Open Road Media, 2014.

[99] Samuel H. Preston. "Family Sizes of Children and Family Sizes of Women." *Demography* 13.1 (1976): 105–14.

[100] William Rhodes, Gerald Gaes, Jeremy Luallen, Ryan King, Tom Rich, and Michael Shively. "Following Incarceration, Most Released Offenders Never Return to Prison." *Crime & Delinquency* 62.8 (2016): 1003–25.

[101] Stuart Robbins. *Lunar Crater Database. Vol. 1*. August 15, 2018. https:// astrogeology.usgs.gov/search/map/Moon/Research/Craters/lunar_crater _database_robbins_2018.

[102] ———. "A New Global Database of Lunar Impact Craters > 1–2 km: 1. Crater Locations and Sizes, Comparisons with Published Databases, and Global Analysis." *Journal of Geophysical Research: Planets* 124.4 (2019): 871–92.

[103] Derek Robertson. "How an Obscure Conservative Theory Became the Trump Era's Go-to Nerd Phrase." *Politico*, February 25, 2018. https://www.politico.com /magazine/story/2018/02/25/overton-window-explained-definition-meaning -217010/.

[104] Oliver Roeder. *Seven Games: A Human History*. W. W. Norton, 2022.

[105] Todd Rose. *The End of Average: How to Succeed in A World That Values Sameness*. Penguin UK, 2016.

[106] Tom Rosentiel. *Four-in-Ten Americans Have Close Friends or Relatives Who Are Gay*. Pew Research Center, May 27, 2007. https://www.pewresearch.org/2007/05 /22/fourinten-americans-have-close-friends-or-relatives-who-are-gay/.

[107] Ryan A. Rossi and Nesreen K. Ahmed. "The Network Data Repository with Interactive Graph Analytics and Visualization." *AAAI*, 2015. http:// networkrepository.com.

[108] Sigal Samuel. *"Should Animals, Plants, and Robots Have the Same Rights as You?"* *Vox*, April 4, 2019. https://www.vox.com/future-perfect/2019/4/4/18285986/robot -animal-nature-expanding-moral-circle-peter-singer.

[109] "SARS-CoV-2 Variants of Concern and Variants Under Investigation In England." Technical Briefing 44, UK Health Security Agency, July 22, 2022. https://assets.publishing.service.gov.uk/government/uploads/system/uploads /attachment_data/file/1093275/covid-technical-briefing-44-22-july-2022.pdf.

[110] Jonathan Schaeffer. *Marion Tinsley: Human Perfection at Checkers?* 2004. https://web.archive.org/web/20220407101006/http://www.wylliedraughts.com/Tinsley.htm.

[111] "Sentences Imposed." Federal Bureau of Prisons, 2019. https://www. bop.gov/about/statistics/statistics_inmate_sentences.jsp.

[112] "September 2005 Aurora Gallery." SpaceWeather.com, 2005. https://spaceweather.com/aurora/gallery_01sep05_page3.htm.

[113] Peter Singer. *The Expanding Circle.* Princeton University Press, 1981.

[114] Charles Percy Snow and Baron Snow. *The Two Cultures and the Scientific Revolution.* Cambridge University Press, 1959.

[115] *Southern California Earthquake Data Center.* Caltech/USGS Southern California Seismic Network, 2022. https://scedc.caltech.edu/.

[116] "Spud Webb." Wikipedia, 2022. https://en.wikipedia.org/wiki/Spud_Webb.

[117] Steven Strogatz. "Friends You Can Count On." *New York Times*, September 17, 2012. https://opinionator.blogs.nytimes.com/2012/09/17/friends-you-can-count-on.

[118] "Supplemental Surveys." United States Census Bureau, 2022. https://www.census.gov/programs-surveys/cps/about/supplemental-surveys.html.

[119] *Surveillance, Epidemiology, and End Results (SEER) Program.* National Cancer Institute. 2016. https://seer.cancer.gov/data/.

[120] *SWPC Data Service.* Space Weather Prediction Center. 2022. https://www.ngdc.noaa.gov/stp/space-weather/solar-data/solar-features/solar-flares/x-rays/goes/xrs/.

[121] Nassim Nicholas Taleb. *The Black Swan: The Impact of the Highly Improbable.* Vol. 2. Random House, 2007.

[122] "The Overton Window of Political Possibility Explained." Mackinac Center. YouTube, February 21, 2020. https://www.youtube.com/watch?v=ArOQF4kadHA.

[123] Derek Thompson. "The Pandemic's Wrongest Man: In a Crowded Field of Wrongness, One Person Stands Out: Alex Berenson." *The Atlantic*, April 1, 2021. https://www.theatlantic.com/ideas/archive/2021/04/pandemics-wrongest-man/618475.

[124] Benjamin Todd. *80,000 Hours: Find a Fulfilling Career That Does Good.* 2022. https://80000hours.org/.

[125] Zeynep Tufekci. "This Overlooked Variable Is the Key to the Pandemic." *The Atlantic*, September 30, 2020. https://www.theatlantic.com/health/archive/2020/09/k-overlooked-variable-driving-pandemic/616548/.

[126] "Tunguska Event." Wikipedia, 2022. https://en.wikipedia.org/wiki/Tunguska_event.

[127] Johan Ugander, Brian Karrer, Lars Backstrom, and Cameron Marlow. "The Anatomy of the Facebook Social Graph." 2011. https://arxiv.org/abs/1111.4503.

[128] Tyler J. VanderWeele. "Commentary. Resolutions of the Birthweight Paradox: Competing Explanations and Analytical Insights." *International Journal of Epidemiology* 43.5 (2014): 1368–73.

[129] James W. Vaupel, Francisco Villavicencio, and Marie-Pier Bergeron-Boucher. "Demographic Perspectives on the Rise of Longevity." *Proceedings of the National Academy of Sciences* 118.9 (2021): e2019536118.

[130] *Vital Statistics Online Data Portal*. Centers for Disease Control and Prevention (CDC), 2018. https://www.cdc.gov/nchs/nsfg.

[131] Allen J. Wilcox and Ian T. Russell. "Perinatal Mortality: Standardizing for Birthweight Is Biased." *American Journal of Epidemiology* 118.6 (1983): 857–64.

[132] Samuel H. Williamson. "Daily Closing Value of the Dow Jones Average, 1885 to Present." MeasuringWorth, 2022. https://www.measuringworth.com /datasets/DJA.

[133] J. Yerushalmy. "The Relationship of Parents' Cigarette Smoking to Outcome of Pregnancy—Implications as to the Problem of Inferring Causation from Observed Associations." *American Journal of Epidemiology* 93.6 (1971): 443–56.

[134] ——. "The Relationship of Parents' Cigarette Smoking to Outcome of Pregnancy—Implications as to the Problem of Inferring Causation from Observed Associations." *International Journal of Epidemiology* 43.5 (2014): 1355–66.

INDEX